THE SUN AND SPACE WEATHER

ASTROPHYSICS AND SPACE SCIENCE LIBRARY

VOLUME 277

EDITORIAL BOARD

Chairman

W.B. BURTON, *National Radio Astronomy Observatory, Charlottesville, Virginia, U.S.A.*
(burton@starband.net); *University of Leiden, The Netherlands* (burton@strw.leidenuniv.nl)

Executive Committee

J. M. E. KUIJPERS, *Faculty of Science, Nijmegen, The Netherlands*
E. P. J. VAN DEN HEUVEL, *Astronomical Institute, University of Amsterdam,*
The Netherlands
H. VAN DER LAAN, *Astronomical Institute, University of Utrecht,*
The Netherlands

MEMBERS

I. APPENZELLER, *Landessternwarte Heidelberg-Königstuhl, Germany*
J. N. BAHCALL, *The Institute for Advanced Study, Princeton, U.S.A.*
F. BERTOLA, *Universitá di Padova, Italy*
J. P. CASSINELLI, *University of Wisconsin, Madison, U.S.A.*
C. J. CESARSKY, *Centre d'Etudes de Saclay, Gif-sur-Yvette Cedex, France*
O. ENGVOLD, *Institute of Theoretical Astrophysics, University of Oslo, Norway*
R. McCRAY, *University of Colorado, JILA, Boulder, U.S.A.*
P. G. MURDIN, *Institute of Astronomy, Cambridge, U.K.*
F. PACINI, *Istituto Astronomia Arcetri, Firenze, Italy*
V. RADHAKRISHNAN, *Raman Research Institute, Bangalore, India*
K. SATO, *School of Science, The University of Tokyo, Japan*
F. H. SHU, *University of California, Berkeley, U.S.A.*
B. V. SOMOV, *Astronomical Institute, Moscow State University, Russia*
R. A. SUNYAEV, *Space Research Institute, Moscow, Russia*
Y. TANAKA, *Institute of Space & Astronautical Science, Kanagawa, Japan*
S. TREMAINE, *CITA, Princeton University, U.S.A.*
N. O. WEISS, *University of Cambridge, U.K.*

THE SUN AND SPACE WEATHER

by

ARNOLD HANSLMEIER

*Institute for Geophysics,
Astrophysics and Meteorology,
University of Graz, Austria*

KLUWER ACADEMIC PUBLISHERS
DORDRECHT / BOSTON / LONDON

A C.I.P. Catalogue record for this book is available from the Library of Congress.

ISBN 1-4020-0684-5

Published by Kluwer Academic Publishers,
P.O. Box 17, 3300 AA Dordrecht, The Netherlands.

Sold and distributed in North, Central and South America
by Kluwer Academic Publishers,
101 Philip Drive, Norwell, MA 02061, U.S.A.

In all other countries, sold and distributed
by Kluwer Academic Publishers,
P.O. Box 322, 3300 AH Dordrecht, The Netherlands.

Printed on acid-free paper

All Rights Reserved
© 2002 Kluwer Academic Publishers
No part of this work may be reproduced, stored in a retrieval system, or transmitted
in any form or by any means, electronic, mechanical, photocopying, microfilming, recording
or otherwise, without written permission from the Publisher, with the exception
of any material supplied specifically for the purpose of being entered
and executed on a computer system, for exclusive use by the purchaser of the work.

Printed in the Netherlands.

Contents

1 Introduction, What is Space Weather?   1

2 The Sun as a Star   3
   2.1 The Sun and the Galaxy 3
     2.1.1 Properties of Stars 4
     2.1.2 Stellar Spectra, the Hertzsprung-Russell-Diagram 5
     2.1.3 Basic Properties of the Sun 9
   2.2 Solar Structure 10
     2.2.1 Hydrostatic Equilibrium 10
     2.2.2 Basic Equations 12
     2.2.3 Energy Generation in the Sun 12
     2.2.4 Convection Zone 13
   2.3 Model: Internal Structure of the Sun 14

3 Observing the Sun   17
   3.1 General Remarks 17
   3.2 Examples of Telescopes 17
   3.3 SOHO 18
   3.4 Solar Polarimetry 20
   3.5 Solar Radio Astronomy 23

4 Phenomenology of Solar Active Regions   25
   4.1 Introduction 25
   4.2 Phenomena in the Solar Photosphere 26
     4.2.1 Radiation Transport 26
     4.2.2 Granulation 27
     4.2.3 Five Minutes Oscillations 32
     4.2.4 Sunspots 33
     4.2.5 Photospheric Faculae 41
   4.3 The Chromosphere 42
     4.3.1 Radiative Transfer in the Chromosphere 43
     4.3.2 Chromospheric Heating 48
     4.3.3 Chromospheric Network, Supergranulation 50
     4.3.4 Solar Flares 51
     4.3.5 Classification of Solar Flares 51

		4.3.6	Where do Flares Occur?	54
	4.4		The Corona	56
	4.5		The Solar Wind	58
		4.5.1	High Speed Solar Wind	62
		4.5.2	Other Diagnostics for the Solar Wind	62
	4.6		Heating of the Corona	64
	4.7		Variations of the Solar Diameter	65
		4.7.1	Satellite Measurements	66
		4.7.2	Measurements with an Astrolabe	67
		4.7.3	Other Semi Diameter Variations	68

5 Testing the Solar Interior — 69
- 5.1 Neutrinos — 69
  - 5.1.1 General Properties — 69
  - 5.1.2 Solar Neutrinos — 70
  - 5.1.3 Solar Neutrino Detectors — 70
  - 5.1.4 Testing the Standard Solar Model — 73
  - 5.1.5 Solution of the Neutrino Problem — 74
- 5.2 Helioseimology-Solar Oscillations — 76
  - 5.2.1 Theory of Solar Oscillations — 79
  - 5.2.2 Helioseismology and Internal Rotation — 82

6 MHD and the Solar Dynamo — 87
- 6.1 Solar Magnetohydrodynamics — 87
  - 6.1.1 Basic equations — 87
  - 6.1.2 Magnetic Buoyancy — 89
  - 6.1.3 Magnetic Flux Freezing — 90
  - 6.1.4 The Induction Equation — 90
  - 6.1.5 Magnetic Reconnection — 92
  - 6.1.6 Fluid Equations — 93
  - 6.1.7 Equation of State — 93
  - 6.1.8 Structured Magnetic Fields — 94
  - 6.1.9 Potential Fields — 95
  - 6.1.10 3 D Reconstruction of Active Regions — 95
  - 6.1.11 Charged Particles in Magnetic Fields — 96
  - 6.1.12 MHD Waves — 98
  - 6.1.13 Magnetic Fields and Convection — 99
- 6.2 The Solar Dynamo — 100
  - 6.2.1 Basic Dynamo Mechanism — 100
- 6.3 Solar Activity Prediction — 107
- 6.4 Stellar Activity — 108

7 The Sun and Climate — 111

- 7.1 The Earth's Atmosphere — 111
  - 7.1.1 Structure of the Atmosphere — 111
  - 7.1.2 Composition — 113
  - 7.1.3 Paleoclimatology — 115
  - 7.1.4 Theory of Milankovich — 119
  - 7.1.5 Greenhouseffect — 121
  - 7.1.6 Ozone — 122
  - 7.1.7 The Structure of the Higher Atmosphere — 123
- 7.2 Earth's History and Origin of the Atmosphere — 125
  - 7.2.1 History of the Earth — 125
  - 7.2.2 Origin of the Atmosphere — 127
- 7.3 The Faint Young Sun — 128
  - 7.3.1 Introduction — 128
  - 7.3.2 The CO_2 Geochemical Cycle — 131
  - 7.3.3 Effects of the Biota — 132
- 7.4 The Atmosphere's Response to Solar Irradiation — 133
  - 7.4.1 Introduction — 133
  - 7.4.2 Solar Variability — 134
  - 7.4.3 Response of the Earth's Atmosphere — 141
  - 7.4.4 Troposphere — 144
  - 7.4.5 Long Term Changes in Solar Irradiance — 145
  - 7.4.6 Solar Protons — 146
- 7.5 Cosmic Rays — 147
  - 7.5.1 The Heliosphere — 148
  - 7.5.2 Cloud, and Cloud Formation Processes — 149
- 7.6 What Causes the Global Warming? — 152

8 Space Weather and Radiation Damage — 155

- 8.1 The Early Sun — 155
  - 8.1.1 T Tauri stars — 155
  - 8.1.2 The Early Sun — 156
- 8.2 Radiation Damage on Living Organisms — 156
  - 8.2.1 Definitions — 156
  - 8.2.2 Radiation Damage on DNA — 158
  - 8.2.3 DNA Repair — 159
  - 8.2.4 Radiation Dose Limits for Astronauts — 159
  - 8.2.5 Genetic vs. Somatic Effects — 160
  - 8.2.6 The Solar Proton Event in August 1972 — 161
- 8.3 Solar UV Radiation Damage — 161
  - 8.3.1 General Remarks — 161
  - 8.3.2 Effects on the Skin — 164
  - 8.3.3 Effects on the Eye — 165
  - 8.3.4 Immune System — 165
  - 8.3.5 UV Index — 165
- 8.4 Spacesuits — 166

		8.4.1 The Extravehicular Mobility Unit 166
	8.5	Radiation Shielding 167

9 The Ionosphere and Space Weather — 169
- 9.1 General Properties 169
  - 9.1.1 Aurora 171
  - 9.1.2 Geomagnetic indices 174
  - 9.1.3 Solar Indices 176
  - 9.1.4 Navigation Systems 177
  - 9.1.5 Radio Communication 178
  - 9.1.6 Geomagnetically Induced Currents 179
  - 9.1.7 Systems Affected by Solar or Geomagnetic Activity 180
- 9.2 Satellites 181
  - 9.2.1 Solar Panels 181
  - 9.2.2 Power Sources for Spacecraft 182
  - 9.2.3 Satellite Crashes 185
  - 9.2.4 Electron Damage to Satellites 186
  - 9.2.5 Single Event Upsets 186
  - 9.2.6 Solar Activity and Satellite Lifetimes 188
  - 9.2.7 The Atmospheric Model 189
  - 9.2.8 Further Reading 191

10 The NOAA Space Weather Scales — 193
- 10.1 Geomagnetic Storms 193
  - 10.1.1 G1 193
  - 10.1.2 G2 194
  - 10.1.3 G3 194
  - 10.1.4 G4 195
  - 10.1.5 G5 195
- 10.2 Solar Radiation Storms 195
  - 10.2.1 S1 195
  - 10.2.2 S2 195
  - 10.2.3 S3 196
  - 10.2.4 S4 196
  - 10.2.5 S5 196
- 10.3 Scale for Radio Blackouts 196
  - 10.3.1 R1 197
  - 10.3.2 R2 197
  - 10.3.3 R3 197
  - 10.3.4 R4 197
  - 10.3.5 R5 197
- 10.4 Summary 198
- 10.5 Space Weather on Mars 198

11 Asteroids, Comets, Meteoroites — **201**
   11.1 Asteroids 201
       11.1.1 General Properties 201
   11.2 Potentially Hazardous Asteroids 202
   11.3 Torino Impact Scale 203
   11.4 NEOs 204
   11.5 The Cretaceous-Tertiary Impact 205
   11.6 Meteorites 208
       11.6.1 The Leonid Threat 210

12 Space Debris — **211**
   12.1 Reentry of Orbital Debris 213
   12.2 Orbital Debris Protection 213
   12.3 ESA Space Debris and Meteoroid Model 215
   12.4 Detection of Space Debris 216
       12.4.1 Radar Measurements 216
       12.4.2 Telescopes 218
       12.4.3 Catalogues 218
       12.4.4 Risk Assessments 218
       12.4.5 Shielding 219
   12.5 Literature 221

13 Appendix — **223**
   13.1 Bibliography 223
   13.2 Internet 236

List of tables — **238**

List of figures — **241**

Chapter 1

Introduction, What is Space Weather?

It is generally accepted that the term *space weather* refers to the time-variable conditions in the space environment that may effect space-borne or ground based technological systems and, in the worst case, endanger human health or life. Therefore there are social and economic aspects of this type of research: one tries to avoid consequences of space weather events by system design or efficient warning and prediction. During the last few years space weather activities have expanded world-wide.

Space weather affects spacecrafts as well as ground based systems.

The main cause for space weather effects is our Sun. It emits a continuous stream of particles which is called the solar wind. The solar wind is variable. It is modulated by the well known 11 year solar activity cycle. Another source of space weather effects are micrometeorites and other space debris.

Related to the solar activity are important effects on spacecraft such as spacecraft charging (surface charging and deep discharges) and single event effects. The effects on humans in space are also to be considered (radiation, particles). Space weather effects also play a rôle on high altitude/high latitude air-flight; cosmic rays penetrate to the lower atmosphere and pose problems to humans and electronic components of modern aeroplanes. Other influences of space weather include radio wave propagation, satellite-ground communications, global satellite-based navigation systems, power transmission systems etc. Changes of the solar irradiance may be one of the causes for climatic changes on the Earth.

Space debris, such as meteoroids, or parts of old satellites must be also be taken into account and are a permanent threat for space missions.

The book is organized as follows. First we want to give a brief review about the main source of space weather effects, our Sun. The basic physics of the Sun will be discussed since it is essential to understand the mechanisms that cause solar variability. This is necessary in order to make prediction models for space weather forecasts. Then we will speak about the influence of solar variability on the Earth's atmosphere. There are similarities with atmospheric weather, however the most important differences between the atmospheric and space weather systems are:

- Meteorological processes are localized; it is possible to make good local weather forecasts. Spaceweather is always global in the planetary scale.

- Space weather events occur over a wide range of time scales: the Earth's magnetosphere responds to solar-originated disturbances within only a few minutes, global reconfiguration occurs within some 10 minutes. Enhanced fluxes of energetic particles in radiation belts decay in time scales of days, months or even longer.

- Spaceweather predictions must rely on the input of just a few isolated measurements of the solar wind and the observations (both ground based and from space) have only a global character sometimes without details.

Therefore, successful space weather activities aiming to make prediction of dangerous events need to be performed on a global scale. Space-borne and ground based observations are complementary.

Presently, the most important users of space weather research are spacecraft engineering, spacecraft operations, RF communications. Spacecraft launchers can make use of exact knowledge of space weather conditions and the re-entry of spacecrafts depends on the atmospheric drag conditions. When the International Space Station is in operation, forecasts will become even more important. Other users are telecommunication operators, users of global positioning systems, electric power industry etc. Commercial airlines must be careful with the radiation doses to their crews and passengers.

I want to thank my students who attended a course on space weather I held at Innsbruck university; they critically read the manuscript and suggested corrections.

Special thanks to Caroline and my children Roland, Christina and Alina; I was allowed to spend lots of nights at the PC. Thanks for the patience of my collaborators because several common projects had to be postponed.

Chapter 2

The Sun as a Star

Our Sun is the only star which is close enough to observe details on its surface. We can observe sunspots, faculae, prominences, coronal holes etc., which are all summarized as solar activity phenomena. Therefore, the study of the Sun is important for astrophysics in general. Theories about stellar structure and evolution can be studied in detail on the Sun.

On the other hand, the Sun is the driving factor for the climate on the Earth and the structure of the Earth's magnetosphere thus determining and influencing the near Earth space environment. Therefore, the study of solar terrestrial relations is of great importance for our modern telecommunication systems both based on Earth and in space. The Sun provides the main input for the so called space weather. Other inputs are meteorites, space debris etc.

2.1 The Sun and the Galaxy

The Sun is a normal star which contains more than 99% of the total mass of the solar system. The solar system, to which the Sun, the 9 great planets, asteroids, meteorites, comets and other small dust particles belong is located in the Milky Way Galaxy. Our galaxy contains more than 2×10^{11} solar masses. The mass of the galaxy can be inferred from the rotation of the system. All stars rotate around the center of the galaxy which is at a distance of about 27 000 light years (Lys) to us (1 Ly $=10^{13}$ km, the distance light travels within one year propagating through space at 300 000 km/s).

At the location of the Sun in the galaxy, one period of revolution about the galactic center is about 200 Million years. Galaxies in general contain some 10^{11} stars, many of them are double systems or may contain planetary systems. The diameter of our galaxy is about 100 000 light years. Galaxies are grouped into clusters- our galaxy belongs to the so called local group of galaxies. The small and large Magellanic cloud are two small dwarf galaxies which are satellites of our system. The nearest large galaxy is the Andromeda galaxy which is at a distance of more than 2 Million light years.

Figure 2.1: A typical spiral galaxy. From a distant galaxy our own system would appear similar, the Sun would be located in one of the spiral arms.

Many galaxies appear as spiral galaxies. Young bright stars are in the spiral arms, older stars in the center and in the halo of the galaxy. An example is given in Fig.2.1.

Thus, from the astrophysical point of view, our sun is situated at about 2/3 from the center of the galaxy and a normal star on a bulk of 10^{11} stars.

2.1.1 Properties of Stars

The only information we can obtain from a star is its radiation. In order to understand the physics of stellar structure, stellar birth and evolution we have to derive quantities such as stellar radii, stellar masses, composition, rotation, magnetic fields etc. We will just very briefly discuss how these parameters can be derived for stars.

- Stellar distances: though this is not an intrinsic parameter for a star, it is of fundamental importance. Stellar distances can be measured by determining their parallax, that is the angle the Earth's orbit would have seen from a star. This defines the astrophysical distance unit *parsec*. A star is at a distance of 1 parsec if the parallax is $1''$. 1 pc = 3.26 Ly.

- Stellar radii: if the apparent diameter of star is known than its real diameter follows from its distance d. The problem is to measure apparent stellar diameters since they are extremely small. One method is to use interferometers, one other method is to use occultation of stars by the moon or mutual occultations of stars in eclipsing binary systems. All these methods are described in ordinary textbooks about astronomy.

- Stellar masses: can be determined by using Kepler's third law in the case we

2.1. THE SUN AND THE GALAXY

observe a binary system. Stellar masses are very critical for stellar evolution, however we know accurate masses only for some 100 stars.

- Once mass and radius are known, the density and the gravitational acceleration follow. These parameters are important for the stellar structure.

- Stellar rotation: For simplicity we can assume that a star consists of two halves, one half approaches to the observer and the spectral lines from that region are blueshifted, the other half moves away and the spectral lines from that area are redshifted. The line profile we observe in a spectrum is a superposition of all these blue- and redshifted profiles and and the rotation causes a broadening of spectral lines;

- Stellar magnetic fields: as it will be discussed in more detail when considering the Sun, magnetically sensitive spectral lines are split into several components when there are strong magnetic fields.

2.1.2 Stellar Spectra, the Hertzsprung-Russell-Diagram

As it has been stressed above, the only information we obtain from stars is their radiation. Putting a prism or a grating inside or in front of a telescope, we obtain a spectrum of a star. Such a spectrum contains many lines, most of them are dark absorption lines. Each chemical element has a characteristic spectrum.

In the Hertzsprung Russell Diagram (HRD) the temperature of stars is plotted versus brightness. The temperature of a star is related to its color: blue stars are hotter than red stars. In the HRD the hottest stars are on the left side. The temperature increases from right to left. Stellar brightness is given in *magnitudes*. The magnitude scale of stars was chosen such that a difference of 5 magnitudes corresponds to a factor of a hundred in brightness. The smaller the number (which can be even negative) the brighter the star. The brightest planet Venus e.g. has magnitude $-4.^m5$ and the Sun has $-26.^m5$. The faintest stars that are visible to the naked eye have magnitude $+6.^m0$. Since the apparent magnitudes depend on the intrinsic luminosity and the distance of a star absolute magnitudes were invented: the absolute magnitude of a star (designated by M) is the magnitude a star would have at a distance of 10 pc. In the HRD we can plot absolute magnitudes as ordinates instead of luminosities. The relation between m and M is given by:

$$m - M = 5\log(r/10) \tag{2.1}$$

r is the distance of the object in pc. For the Sun we have $M = +4.^M5$; from a distance of 10 pc it would be among the fainter stars visible with the naked eye.

How can we determine stellar temperatures? Stars can be considered to a very good approximation as *black body* radiators. A black body is a theoretical idealization: an object that absorbs completely all radiation at all wavelengths. The radiation of a black body at a given temperature is given by the *Planck law*:

$$I_\nu = B_\nu = (2h\nu^3/c^2)/\exp(h\nu/kT_S) - 1 \tag{2.2}$$

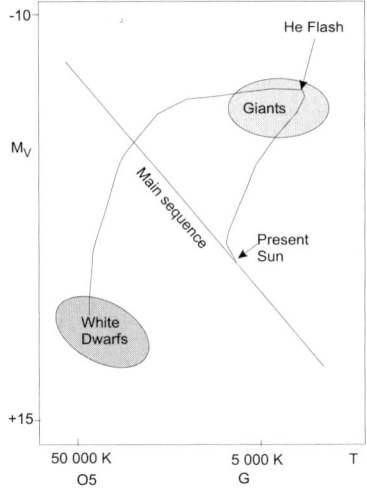

Figure 2.2: Sketch of the Hertzsprung-Russell-diagram with evolutionary path of the Sun.

thus it depends only on the temperature T_S of the object. Here, I_ν is the intensity of radiation at frequency ν; h, k, c are Planck's constant, Boltzmann's constant and the speed of light. $h = 6.62 \times 10^{-34}\,\text{Js}^{-1}$, $k = 1.38 \times 10^{-23}\,\text{JK}^{-1}$. If that equation is integrated over all frequencies (wavelengths), we obtain a formula for the total power emitted by a black body:

$$\int_0^\infty B_\lambda d\lambda = \sigma T^4, \tag{2.3}$$

and for the luminosity of a star:

$$L = 4\pi r^2 \sigma T_{\text{eff}}^4 \tag{2.4}$$

For the Sun $T_{\text{eff}} = 5\,785\,\text{K}$. This formula defines the effective temperature of a star. $\sigma = 5.67 \times 10^{-8}\,\text{W/m}^2\text{K}^4$ is the Stefan Boltzmann constant.

What is the power emitted per unit area of the Sun's surface? Answer: Put T = 6 000 K we find that the Sun radiates 70 MW per m² of its surface.

By taking the derivative with respect to λ of Planck's Law and setting it equal to zero, one can find the peak wavelength, where the intensity is at maximum:

$$T\lambda_{\text{max}} = 2.9 \times 10^{-3}\,\text{m K} \tag{2.5}$$

This is also called *Wien's law*.

At about what wavelength can planets be expected to radiate most of their energy?

2.1. THE SUN AND THE GALAXY

Table 2.1: Central wavelengths and bandwidth of the UBVRI filter set

Name	Meaning	Central λ	Bandwidth [nm]
U	Ultraviolet	360	66
B	Blue	440	98
V	Visual (green)	550	87
R	Red	700	207
I	Infrared	900	231

Answer: Let us assume the temperature of the Earth = 300 K. Then

$$\lambda_{max} = 2.9 \times 10^{-3}/300 \sim 10\,\mu \tag{2.6}$$

The Sun has a surface temperature of about 6 000 K. At what wavelength does the Sun's spectrum peak? Answer:

$$\lambda_{max} = 2.9 \times 10^{-3}/6000 \sim 0.5\,\mu = 500\,\text{nm} \tag{2.7}$$

Therefore, we can measure from the spectrum directly the temperature of stars. The temperature derived from the peak wavelength is called Wien Temperature, the temperature derived from the difference between two intensity levels (=color) Color temperature etc. In order to define color, a filter system must be defined. The most commonly used system is the UBV system which has three bands that are located in the UV (U), blue (B) and visual (V) to measure the intensity I_ν. The luminosity of stars is given in magnitudes which are defined as follows:

$$\text{Magnitude} = \text{const} - 2.5\log(\text{Intensity}) \tag{2.8}$$

The color of a star is measured by comparing its magnitude through one filter (e.g. red) with its magnitude through another (e.g. blue).

E.g. m_V means the magnitude measured with the V filter.

Therefore, instead of determining temperatures from the comparison of the spectrum of a star with the Planck law, one can use e.g. color indices: Let us compare a hot star (ζPup) with a cool star (Betelgeuze). If we calculate B-V, than this value will be:

- positive for the cooler star, since it is brighter in V than in B (blue). If the cool star is brighter in V it means that its magnitude has a lower value and therefore B-V is positive.

- negative for the hotter star. The hotter star is brighter in B than in V, therefore for the magnitudes in these two bands: $m_B < m_V$ and B-V<0.

The HRD consists of:

Table 2.2: B-V colors and effective temperatures of some stars

Star	B-V	Effective T
Sun	+0.6	5 800 K
Vega	0.0	10 000 K
Spica	-0.2	23 000 K
Antares	+1.8	3 400 K

- Main sequence stars: they lie on a diagonal from the upper left (hot) to the lower right (cool).

- giants, supergiants: they have the same temperature as the corresponding main sequence stars but are much brighter and must have larger diameters (see equation 2.4).

- White dwarfs are faint but very hot objects thus from their location at the lower left in the HRD it follows that they must be very compact (about 1/100 the size of the Sun). In about 5×10^9 years our Sun will evolve into a red giant and finally to a white dwarf.

The evolutionary path of our Sun is also given in the 2.2. We have the following main evolutionary steps:

- Pre main sequence evolution: from a protostellar gas and dust cloud the Sun was formed and before it reaches the main sequence where it spends most of its life, the contracting Sun passes a violent youth, the T Tauri phase.

- At the main sequence the Sun changes extremely slowly and remains about 10^{10} years there. In the core H is transformed to He by nuclear fusion.

- The Sun evolves to a red giant, it will expand and the Earth will become part of the solar atmosphere. The expansion starts when all H is transformed to He in the core. Then a H burning shell supplies the energy. The He flash sets in as soon as in center the He burning sets in. The Sun will be a red giant for some 10^8 ys.

- Finally the Sun becomes a white dwarf which slowly cools.

In terms of radius we have:

- present Sun: 1 $R_\odot$,

- red giant: $\sim 10^4 R_\odot$,

- white dwarf: $0.01 R_\odot$.

In terms of space weather these evolutionary effects are negligible. But it is interesting to investigate them especially for the early Sun.

Table 2.3: Spectral classification of stars

O	ionized He, ionized metals
B	neutral He, H stronger
A	Balmer lines of H dominate
F	H becomes weaker, neutral and singly ionized metals
G	singly ionized Ca, H weaker, neutral metals
K	neutral metals molecular bands appear
M	TiO, neutral metals
R,N	CN, CH, neutral metals
S	Zirconium oxide, neutral metals

Table 2.4: Effective Temperature as a function of spectral type

Spectral Type	O	B0	A0	F0	G0	K0	M0	M5
T_{eff} [K]	50 000	25 000	11 000	7 600	6 000	5 100	3 600	3 000

For the main sequence stars there exists a relation between their mass and luminosity:

$$L \sim M^4 \qquad (2.9)$$

From 2.9 we see that more massive stars are very luminous and therefore they use up their nuclear fuel much more rapidly than low massive stars like our Sun. Massive main sequence stars that are observed today must have been formed in very recent astronomical history.

According to their spectra, stars can be classified in the following sequence: O-B-A-F-G-K-M. This is a sequence of temperature (see Table 2.4): O stars are hottest, M stars coolest; the number of absorption lines increase from O to M. Some characteristics are given in Table 2.3.

The luminosity of a star depends on a) temperature, b) surface which is $\sim R^2$. Since e. g. a K star may be a dwarf main sequence star or a giant luminosity classes have been introduced. Class I contains the most luminous supergiants, class II the less luminous supergiants. Class III are the normal giants, Class IV the sub giants and class V the main sequence.

Now we understand the spectral classification of our Sun: it is a G2V star.

2.1.3 Basic Properties of the Sun

As it has been mentioned, the Sun is a G2V star in the disk of our Galaxy. The mass of the Sun is:

$$M_\odot = 1.99 \times 10^{30} \, \text{kg} \qquad (2.10)$$

An application of Kepler's third law gives us the mass of the Sun if its distance is known which again can be derived from Kepler's third law:

$$\frac{a^3}{P^2} = \frac{G}{4\pi^2}(M_1 + M_2) \qquad (2.11)$$

In our case a denotes the distance Earth-Sun (150×10^6 km), P the revolution period of the Earth around the Sun (1 year), M_1 the mass of the Earth and M_2 the mass of the Sun. One can make the assumption that $M_1 \ll M_2$ and therefore $M_1 + M_2 \sim M_2$.

If we know the distance of the Sun and its angular diameter the solar radius is obtained:

$$r_\odot = 6.96 \times 10^8 \text{ m} \qquad (2.12)$$

In order to determine it, we must use the Sun's distance and make a measurement of the angular diameter of the visible solar disk. This is not quite easy; one possibility is to define the angular distance between the inflection points of the intensity profiles at two opposite limbs. Such profiles can be obtained photoelectrically and the apparent semidiameter at mean solar distance is about 960 seconds of arc ('').

Knowing the mass and radius of the Sun, the mean density can be calculated:

$$\bar{\rho} = 1.4 \text{ g/cm}^3 \qquad (2.13)$$

The gravitational acceleration is given by:

$$g = GM/R^2 = 274 \text{ m/s}^2 \qquad (2.14)$$

The *solar constant* is the energy crossing unit area of the Earth's surface perpendicular to the direction from the Earth to the Sun in every second. In SI the units are W m^{-2}. UV and IR radiation from the Sun is strongly absorbed by the Earth's atmosphere. Therefore, accurate measurements of the solar constant are only possible with satellites. ACRIM on SMM and ERB on Nimbus 7 showed clearly that the presence of several large sunspots which are cooler than their surroundings depress the solar luminosity by $\sim 0.1\%$. The solar luminosity is:

$$L_\odot = 3.83 \times 10^{26} \text{ W} \qquad (2.15)$$

And the effective temperature:

$$T_{\text{eff}\odot} = 5780 \text{ K} \qquad (2.16)$$

2.2 Solar Structure

2.2.1 Hydrostatic Equilibrium

How a ball of gas and plasma, like a star remains stable against gravitational collapse or free expansion? Let us assume a sphere of mass M and radius R. In most cases there are only two forces:

2.2. SOLAR STRUCTURE

- gravity: acts inward
- pressure: acts outward

Let us consider a shell inside a star, the lower boundary is at r from the center and the upper at $r + \Delta r$. ΔA is a surface element and P_{outer}, P_{inner} denote the pressure at r and $r + \Delta r$. The net force on such a shell is:

$$F_{net} = F_{grav} - F_p \qquad (2.17)$$

and $F_p = (P_{outer} - P_{inner})\Delta A$. From the above equation:

$$F_p = [P(r) + (dP/dr)\Delta r - P(r)]\Delta A = (dP/dr)\Delta r \Delta A \qquad (2.18)$$

By dividing the net force F_{net} by $-\Delta m = -\rho(r)\Delta r \Delta A$, we find the equation of motion of the shell:

$$-d^2r/dt^2 = g(r) + [1/\rho(r)](dP/dr) \qquad (2.19)$$

If the acceleration is set to zero (when there is a balance), then the *hydrostatic equilibrium* becomes:

$$\frac{dP}{dr} = -\frac{GM(r)\rho(r)}{r^2} \qquad (2.20)$$

Therefore, the pressure at depth h must be high enough to support the weight of the fluid per unit area above that depth. Let us derive an estimate for the central pressure of a star. The pressure is given by:

$$P = g\rho h \qquad (2.21)$$

At the center $h = R$; from $g = GM/R^2$ we find the central pressure P_c:

$$P_c = \frac{GM\rho}{R} \qquad \rho = \frac{M}{4\pi R^3/3} \qquad (2.22)$$

which leads to:

$$P_c = \frac{3}{4\pi}\frac{GM^2}{R^4} \qquad (2.23)$$

For the Sun: $M_\odot = 2\times 10^{30}$ kg, and $R_\odot = 7\times 10^8$ m. This gives $P_c = 3\times 10^{14}$ Nm^{-2} compared to the atmospheric pressure at sea level on Earth of 10^5 Nm^{-2}. This is a very crude approximation, since in reality the density increases with depth and the true central pressure of the Sun is 100 times larger than the estimate.

Compare the central pressure of a main sequence star of $M = 10\,M_\odot$ and $R = 4\,R_\odot$ to the central pressure of the Sun. Answer: from $P \sim M^2 R^{-4}$ the central pressure is $10^2 \times 4^{-4} = 0.4$ times the central pressure of the Sun.

What happens if a star contracts (which will be the case when the hydrostatic equilibrium condition is not established)? According to the *Virial Theorem* half of the gravitational energy which is set free is radiated away and the other half heats the star.

2.2.2 Basic Equations

In most phases of stellar evolution, the structure of a star can be determined by the solution of four first order differential equations:

- hydrostatic equilibrium,
- mass continuity,
- gradient of luminosity (related to energy release),
- temperature gradient.

The corresponding equations are:

$$\frac{dP}{dr} = -\frac{GM\rho}{r^2} \quad (2.24)$$

$$\frac{dM}{dr} = 4\pi r^2 \rho \quad (2.25)$$

$$\frac{dL}{dr} = 4\pi r^2 \rho \epsilon \quad (2.26)$$

$$\frac{dT}{dr} = -\frac{3\kappa L\rho}{16\pi a c r^2 T^3} \quad (2.27)$$

In these equations r is the distance from the stellar center, P, ρ, T are the pressure, density and temperature at radius r, M is the mass contained within r, L the energy carried by radiation across r, ϵ the nuclear energy release. The quantities P, ϵ, κ depend on density, temperature and composition. κ is the *opacity* and measures the resistance of the material to energy transport.

2.2.3 Energy Generation in the Sun

In principle, a variety of different energy generating processes can take place in stars. When the star first forms from the interstellar medium it contracts, radiating away gravitational energy (see the Virial Theorem). During this stage no nuclear reactions take place and half of the gravitational energy is radiated away, the other half increases the temperature of the core. As soon as the central temperature exceeds about 10^6 K nuclear reactions start. Energy is generated by the fusion of two lighter particles to form a heavier particle whose mass is smaller than the mass of its constituents, the mass defect being transformed into energy according to $E = \Delta M c^2$.

Let us consider the fusion of H into He. The mass of 4 H is:

$$4 \times 1.008145 \,\text{AMU} \quad (2.28)$$

The mass of the resulting He atom is

$$4.00387 \,\text{AMU} \quad (2.29)$$

Thus the difference is:

$$0.02871 \,\text{AMU} \sim 4.768 \times 10^{-26} \,\text{g} \sim 4.288 \times 10^{-5} \,\text{erg} \sim 26.72 \,\text{MeV} \quad (2.30)$$

2.2. SOLAR STRUCTURE

Table 2.5: The principal reaction of the pp chain

Reaction Number	Reaction	Neutrino Energie (MeV)
1	$p + p \rightarrow {}^2H + e^+ + \nu_e$	0.0 to 0.4
2	$p + e^- + p \rightarrow {}^2H + \nu_e$	1.4
3	${}^2H + p \rightarrow {}^3He + \gamma$	
4a	${}^3He + {}^3He \rightarrow {}^4He + 2p$	
4b	${}^3He + {}^4He \rightarrow {}^7Be + \gamma$	
5	$e^- + {}^7Be \rightarrow {}^7Li + \nu_e$	0.86, 0.38
6a	${}^7Li + p \rightarrow {}^4He + {}^4He$	
6b	$p + {}^7Be \rightarrow {}^8B + \gamma$	
7	${}^8B \rightarrow {}^8Be + e^+ + \nu_e$	0...15

If one assumes that the Sun consists of pure hydrogen which is converted into He, then the total energy would be 1.27×10^{52} erg. The luminosity of the Sun is $L_\odot = 3.8 \times 10^{33}$ erg/s thus there would be energy supply for 10^{11} years.

Let us consider the conversion of H into He in more detail. The so called pp chain (Table 2.5) dominates in stars with relatively low central temperatures (like the Sun) and the CN cycle is important for stars at higher central temperatures.

Let us consider the principal reactions for the pp cycle. If reaction 4b is realized then the further reactions can take place. If reaction 6b occurs, then reaction 7 follows.

The energy production rate ϵ depends highly on the temperature:

$$\epsilon \sim \rho T^5 \qquad (2.31)$$

The pp reaction dominates for stars with central temperatures between 5 and 15×10^6 K.

The CN cycle also produces He from H where C is used as a catalyst.

2.2.4 Convection Zone

In the solar core nuclear fusion generates the energy which is transported outwards by radiation. At a depth of a third of the solar radius below the solar surface the convection zone starts, where energy is transported outwards by convective motions. This zone occupies only 2% of the solar mass. Hydrogen and He are neutral at the solar surface but they are ionized just below the surface. In these ionization zones the ratio of the specific heat at constant pressure (c_p) to the specific heat at constant volume (c_v) is much lower than the value 5/3. This value is appropriate either to a neutral gas or fully ionized gas. Because $c_p/c_v \ll 5/3$

convection occurs. For the temperature gradient we already have seen that:

$$\frac{dT}{dr} = -\frac{3\kappa L_{\rm rad}\rho}{16\pi acr^2 T^3} \tag{2.32}$$

The total luminosity L is:

$$L = L_{\rm rad} + L_{\rm conv} \qquad L_{\rm conv} = ? \tag{2.33}$$

Basically, convection can be treated as an instability; if an element of material is displaced upwards, then it continues to rise if it is lighter than its surroundings. By assuming that the rising element moves sufficiently slowly that it is in pressure balance with its surroundings but that at the same time its motion is adiabatic (no heat exchange between the element and the surroundings), then convection occurs if:

$$\frac{P}{\rho}\frac{d\rho}{dP} < \left(\frac{P}{\rho}\frac{d\rho}{dP}\right)_{\rm ad} \tag{2.34}$$

If the stellar material is an ideal classical gas with constant ratio of specific heats γ, then:

$$\frac{P}{\rho}\frac{d\rho}{dP} < \frac{1}{\gamma} \tag{2.35}$$

The theory which is usually used contains a free parameter, the so called *mixing length* l:

$$l = \alpha H_{\rm p} = \alpha \left| P\frac{dP}{dr} \right| \tag{2.36}$$

where H_P is the pressure scale height. It is supposed that α is order of unity. As we will discuss later, information about the depth of the convection zone comes from a detailed study of solar oscillations. Apart from energy transport one has also to consider that in convection zones there is a uniform chemical composition. This prevents any attempt of heavy chemical elements to settle in the Sun's gravitational field (any stratification produced by radiation pressure).

2.3 Model: Internal Structure of the Sun

In this paragraph we give a table showing the variation of temperature, luminosity and fusion rate as a function of increasing distance from the solar center. Such a model can be calculated from the basic set of equations discussed above.

Solar models computed with mass loss, microscopic diffusion of helium and heavy element, and with updated physics have been evolved from the pre-main sequence to present day (Morel et al., 1997); they are compared to the observational constraints including lithium depletion and to the seismic reference model of Basu et al. (1996), derived by inversion. Microscopic diffusion significantly improves the agreement with the observed solar frequencies and agree with the seismic reference model within ±0.2% for the sound velocity and ±1% for the density, but slightly worsens the neutrino problem.

2.3. MODEL: INTERNAL STRUCTURE OF THE SUN

Table 2.6: Solar model: variation of temperature, luminosity and fusion rate throughout the Sun

% Radius	Radius [10^9] m	Temperature [10^6] K	% Luminosity	Fusion rate [J/kg s]
0	0.00	15.7	0	0.0175
0.09	0.06	13.8	33	0.010
0.12	0.08	12.8	55	.0068
0.14	0.10	11.3	79	.0033
0.19	0.13	10.1	91	.0016
0.22	0.15	9.0	97	0.0007
0.24	0.17	8.1	99	0.0003
0.29	0.20	7.1	100	0.00006
0.46	0.32	3.9	100	0
0.69	0.48	1.73	100	0
0.89	0.62	0.66	100	0

Chapter 3

Observing the Sun

3.1 General Remarks

In this short chapter we want to give a few examples of modern solar telescopes. Some remarks are also given concerning optical design and features as well as disturbances caused by the Earth's atmosphere.

Several factors influence the image quality of solar telescopes. Sunlight can heat up the telescope structure and the main optics causing the so called internal seeing. Considering reflecting telescopes, in particular the main mirror absorbs up to 10% of the collected light and its surface may heat up considerably leading to mirror seeing. The most effective measure to prevent internal seeing is to remove the air entirely, the telescope is evacuated. A window at the entrance and exit preserves the vacuum. The main problem here is to have a window with high optical quality which is thick enough to resist air pressure. Helium filling is an alternative to evacuation. The viscosity of He and the dependence of the index of refraction from temperature are lower than for air whereas temperature conductivity is higher. A forced flow of He inside the telescope tube cancels inhomogeneities. The THEMIS telescope (Telescope Heliographique pour l'etude du Magnetisme et des Instabilites Solaires, see Arnaud et al. 1998, Mein et al., 1997) has a He filled tube. Other possibilities are to construct open telescopes, such as the DOT (Dutch open telescope at La Palma, see Rutten et al., 2000).

To reduce atmospheric turbulence systems which dynamically control the wavefront deformations effected by the atmosphere are used (adaptive optics). For THEMIS a joint between the telescope tube and the dome which has an entrance window of 1 m prevents air exchanges between outside and inside the dome.

3.2 Examples of Telescopes

Very briefly some selected examples are given. The Big Bear Solar Observatory (BBSO) (Fig. 3.1) is located at 2000 m elevation in the middle of Big Bear Lake. This site reduces the image distortion which usually occurs when the Sun heats the

Figure 3.1: Big Bear Solar Observatory

ground and produces convection in the air just above ground. Turbulent motions in the air near the observatory are also reduced by the smooth flow of wind across the lake instead of turbulent flow that occurs over mountain peaks and forests. The main instrument is a 65 cm reflector.

In Fig. 3.2 a drawing of a solar vacuum tower telescope is given. Light enters the vacuum tank through a coelostat system and a mirror. The vertical tank is evacuated in order to avoid turbulence in the telescope itself. At NSO, Kitt Peak, the telescope is a 70 cm f/52 system.

The German Vacuum Tower Telescope (VTT) at the Observatorio del Teide, Tenerife has two coelostat mirrors (80 cm) and the entrance window to the vacuum tank (BK7) has a diameter of 75 cm and a thickness of 7 cm. The primary mirror has 70 cm and the focal length of the system is 45.64 m. The total field of view is 700 arcsec and the scale is 4.52 arcsec/mm.

Other famous solar instruments for observing the Sun in high spatial resolution mode are the Coupole at the Observatoire Pic du Midi and the Swedish La Palma Solar Telescope.

A recent review about solar instrumentation was given by v.d. Lühe (2001).

3.3 SOHO

The **SO**lar and **H**eliospheric **O**bservatory is a common project being carried out by the European Space Agency (ESA) and the US National Aeronautics and Space Administration (NASA) in the framework of the Solar Terrestrial Science Program (STSP) comprising other missions like CLUSTER and the International Solar Terrestrial Physics Program (ISTP) with Geotail, WIND and Polar. SOHO was launched on December 2, 1995.

SOHO is located at the Lagrangian point L1 about 1.5 Million km away from Earth which permits an uninterrupted view of the Sun. All previous space solar ob-

3.3. SOHO

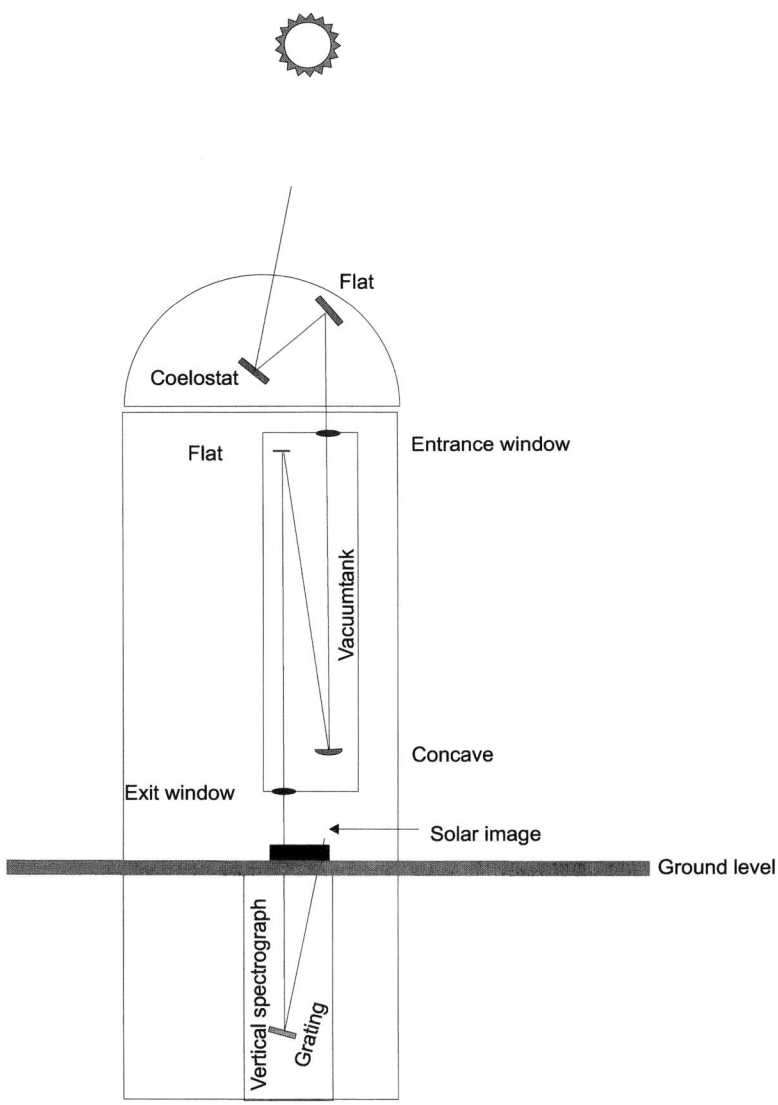

Figure 3.2: Optical path scheme of a vacuum telescope (e.g. Kitt Peak or VTT, Tenerife). Below the ground level, a vertical spectrograph is located. The solar image can be observed at the top of the optical bank that is shown as a black box in the sketch.

servatories have orbited the Earth, from where their observations were periodically interrupted as our planet 'eclipsed' the Sun.

The main scientific objectives of SOHO are:

- Interior of the Sun
- Solar atmosphere
- Solar wind

The main instruments are:

- CDS: Coronal Diagnostic Spectrometer,
- CELIAS: Charge, Element, and Isotope Analysis System,
- COSTEP: Comprehensive Suprathermal and Energetic Particle Analyzer,
- EIT: Extreme ultraviolet Imaging Telescope,
- ERNE: Energetic and Relativistic Nuclei and Electron,
- GOLF: Global Oscillations at Low Frequencies,
- LASCO: Large Angle and Spectrometric Coronagraph Experiment,
- MDI: The Michelson Doppler Imager,
- SUMER: Solar Ultraviolet Measurements of Emitted Radiation
- SWAN: Solar Wind ANisotropies,
- UVCS: UltraViolet Coronagraph Spectrometer,
- VIRGO: Variability of solar IRradiance and Gravity Oscillations.

Further details about SOHO and the instruments can be found e.g. in the review given by Fleck (2001).

A drawing of the spacecraft is given in Fig. 3.3.

3.4 Solar Polarimetry

As we will see, magnetic fields are the key to understand solar activity phenomena and to predict them. To measure magnetic fields on the Sun, it is inevitable to deal with polarimetry.

Electromagnetic radiation consists of oscillations of electric and magnetic fields perpendicular to the direction of propagation. The electric field vector determines the polarization of an electromagnetic wave:

- circular polarization: $\vec{E}$ rotates with its endpoint describing a circle in the plane of polarization, right or left handed, depending on the sense of rotation.

3.4. SOLAR POLARIMETRY

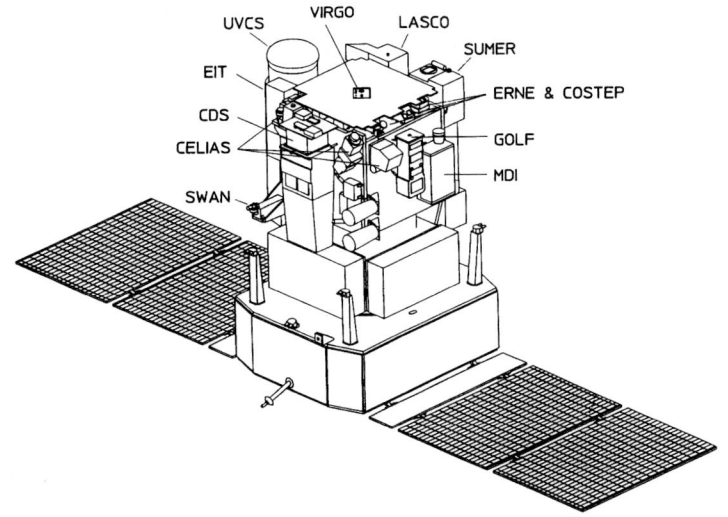

Figure 3.3: Drawing of the SOHO solar observatory (ESA & NASA)

- linear polarization: $\vec{E}$ remains in a fixed position

- unpolarized: orientation of $\vec{E}$ changes randomly with time.

Polarization can be mathematically described by the four Stokes Parameters

- I intensity,

- Q the linear polarization in the direction of the position angle 0^0,

- U the linear polarization at 45^0,

- V the circular polarization. For positive V, the vector $\vec{E}$ is rotating clockwise as seen from the observer.

Usually the polarization parameters are given relative to the intensity, i.e. Q/I, U/I, and V/I.

A photodetector can be used to measure the intensity of light generating an electric current proportional to the intensity I. The polarization parameters however cannot be measured directly. They have to be measured by using optical devices which are polarization sensitive (linear polarizers, retarders). In a polarimeter, the Stokes vector $S = (I, Q, U, V)$ is transformed into $S' = (I', Q', U', V')$, where the transmitted intensity I' depends on Q, U, V.

Let us consider a simple example: The intensity I of linearly polarized light is measured with a photodetector. After introducing a linear polarizer into the light beam with arbitrary orientation the measured intensity becomes $I' < I$.

The light is unpolarized if for all orientations of the polarizer the intensity is the same. The orientation of the linear polarizer is varied and thus the intensity of the light measured with the photodetector is modulated.

Because of seeing effects, for precise polarization measurements the modulation frequency must lie above the frequency of the intensity fluctuations caused by disturbances e.g. atmospheric turbulences in solar observations.

The action of linear optical systems on polarized light is described by the Mueller matrices:

$$\vec{I'} = \mathbf{M}\vec{I} \tag{3.1}$$

The intensity of the outcoming beam is a linear combination of all four Stokes parameters of the incoming beam. The Stokes parameters must verify the following conditions:

$$I^2 - Q^2 - U^2 - V^2 \geq 0 \qquad I \geq 0 \tag{3.2}$$

When doing solar polarimetry one has to take into account for different effects:

- environmental polarization: a typical observation includes the Earth's atmosphere $\mathbf{M_A}$, the telescope $\mathbf{M_T}$, the polarimeter $\mathbf{M_P}$, the spectrograph $\mathbf{M_S}$ and the detector $\mathbf{M_D}$. As light passes these systems, the final Mueller matrix is given by:

$$\mathbf{M} = \mathbf{M_D}\mathbf{M_S}\mathbf{M_P}\mathbf{M_T}\mathbf{M_A} \tag{3.3}$$

The Stokes parameters $\vec{I}_{\text{sun}}$ are related to the observed Stokes parameters by

$$\vec{I}_{\text{sun}} = \mathbf{M}^{-1}\vec{I}_{\text{obs}} \tag{3.4}$$

The Earth's atmosphere is not static, time fluctuations produce wavefront distortions that lead to spatial smearing of the image and spurious polarization features. When plotting a power spectrum of the polarimetric signal it is seen that the individual distributions are broadened and overlap (cross talk).

- instrumental polarization: the air inside a telescope can be turbulent and produce similar effects to those of the atmosphere. That can be avoided by keeping the telescope in vacuum or some other means. Metallic surfaces act as partial polarizers and retarders. Windows made of glass can have mechanic tensions and stresses producing inhomogeneous birefringence and thus they behave as retarders.

- spectrograph and detector polarization

The two basic effects that can be used for measuring magnetic fields are:

- Zeeman effect: degeneracy of the atomic eigenstates effected by the magnetic field, splitting of line profiles and characteristic polarization (Stokes V circular, Stokes Q linear). Can be used for magnetic fields $> 100\,\Gamma$ to compete with the microturbulent Doppler broadening of the line profiles.

3.5. SOLAR RADIO ASTRONOMY

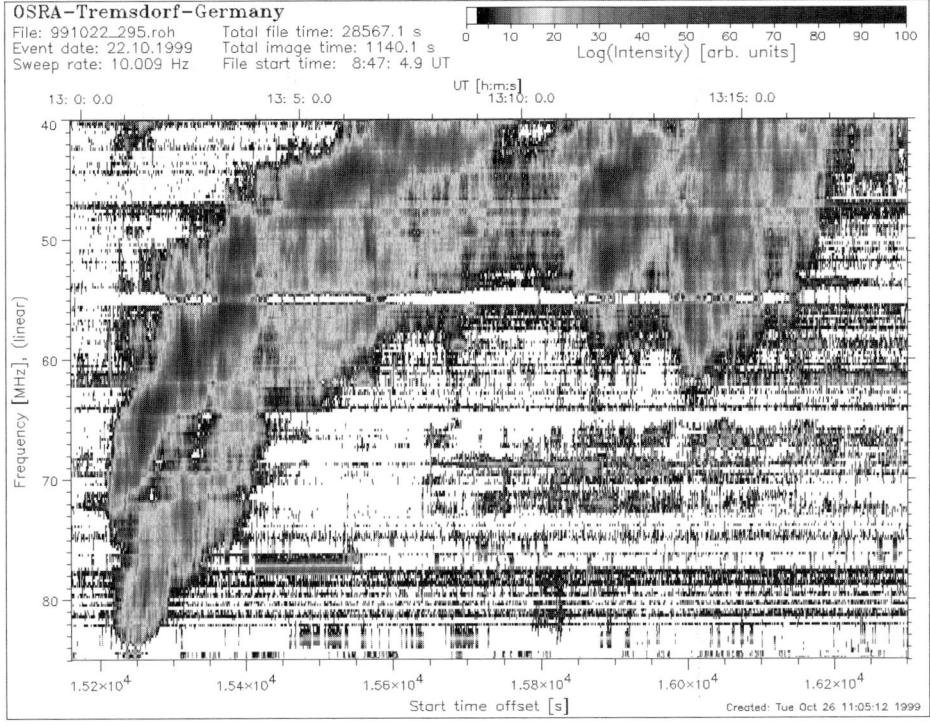

Figure 3.4: Propagation of a wave throughout the outer solar atmosphere. On the abscissa is the time, on the ordinate the frequency. Within 5 min the frequency drifts from 80 MHz to 40 MHz indicating the propagation to the higher corona. Courtesy: H. Aurass, Th. Mann, AIP.

- Hanle effect: useful diagnostic where the magnetic field is relatively weak (a few to a few tens of Gauss) and where the plasma is sufficiently tenuous that collisional excitation can be neglected in comparison to the radiative excitation of the upper level. It introduces both a rotation of the plane of polarization and a reduction of the net polarization of the scattered light.

3.5 Solar Radio Astronomy

The solar corona is an inhomogeneous, hot, dilute and fully ionized plasma. Its spatial structure is governed by the magnetic field. Plasma processes associated with solar activity take place on small temporal and spatial scales and reveal themselves by electron acceleration (up to a few MeV) which can emit radio radiation. These are called non thermal electrons. Radio radiation from the Sun was first detected in 1942.

From a simple derivation from electrodynamics it follows, that there exists a plasma frequency, ω_P and that propagating waves through such a plasma are only possible for $\omega_{\text{waves}} < \omega_P$

$$\omega_P = \sqrt{\frac{4\pi n e^2}{m_e}} \qquad (3.5)$$

where n is the number of electrons per unit volume and ω_P the *plasma frequency*. Let us consider the propagation of an electromagnetic wave throughout the Sun's outer layers. Here, the density decreases from the chromosphere to the corona. Therefore, electromagnetic waves with higher frequencies originate in deeper layers and by observing the Sun in different frequency channels one can measure the propagation of a plasma wave through the atmosphere (see Fig. 3.4).

Possible emission mechanisms at cm- and mm-wavelengths are for the active Sun component:

The quiet Sun component of the radio emission is from thermal emission from the hot ionized gas.

At a frequency of 100 GHz (0.3 cm), the emission originates at the same height in the photosphere than at visible wavelengths. At 1.4 GHz (21 cm) the emission originates from the top of the chromosphere (corresponding to a black body at 100 000 K) and at longer wavelengths (e.g. 300 cm, corresponding to 0.1 GHz) the emission arises from the corona (1-2 Million K blackbody). Thus, the size of the Sun varies when measured at different wavelengths.

The slowly varying component also has thermal origin and arises from regions above sunspots where the electron density is higher

Chapter 4

Phenomenology of Solar Active Regions

4.1 Introduction

The Sun and its atmosphere can be divided as follows:

- Solar interior: can be further subdivided into

  1. Core: about 1/3 of the solar radius; here energy production occurs.
  2. Radiation zone: about 1/3 of the solar radius; the energy is transported outward by innumerable emission and absorption processes transferring the high energy γ photons that are produced by nuclear fusion into longer wave photons.
  3. Convection zone: starts below the surface extending about 2×10^5 km into the interior.

- Solar atmosphere: can be subdivided into

  1. Photosphere: starts at the surface (which can be defined as the region where light is absorbed considerably over a short distance) and extends up to 500 km.
  2. Chromosphere: above the photosphere; extends to about 2 Mm.
  3. Transition Region: strong increase of temperature up to 10^6 K over a very small spatial range (some 10^4 km).
  4. Corona: starts above 2 Mm, high temperature $> 10^6$ K.

In Fig. 4.1 the variation of temperature and electron density is shown.

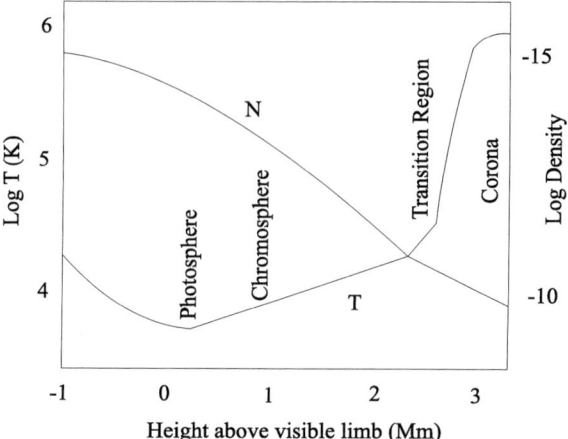

Figure 4.1: Variation of electron temperature and electron density in the solar atmosphere

4.2 Phenomena in the Solar Photosphere

4.2.1 Radiation Transport

The photosphere of the Sun (or of a star) is the layer which can be seen in the visible portion of the continuous radiation spectrum. Here, the photons in the continuum of the visible spectrum have their last scattering encounter before leaving the atmosphere. Let the opacity κ_ν be that fraction of a beam of radiation of frequency ν and intensity I_ν which is absorbed or scattered out of the beam per unit distance. The scattering occurs by atoms, molecules or electrons of the plasma through which it passes. Let us define for an element of plasma of thickness dz and opacity $\kappa_\nu(z)$ the *optical thickness* $d\tau$ by:

$$d\tau = -\kappa_\nu(z)dz \tag{4.1}$$

and hence

$$\tau_\nu(z) = -\int_0^z \kappa_\nu(z)dz \tag{4.2}$$

The transfer of radiation through the atmosphere of a star is governed by the equation of radiative transfer. If θ denotes the angle between the direction of the beam of radiation and the outward normal, and $\mu = \cos\theta$ then under the assumptions that a) the atmosphere is plane - parallel and b) is locally in thermodynamic equilibrium (LTE), the transport equation becomes:

$$\mu \frac{\partial I_\nu(\tau_\nu,\mu)}{\partial \tau_\nu} = B_\nu(T) - I_\nu(\tau_\nu,\mu) \tag{4.3}$$

where $B_\nu(T)$ is the Planck function at temperature T:

$$B_\nu(T) = \frac{2h\nu^3}{c^2}\left(e^{h\nu/kT} - 1\right)^{-1} \tag{4.4}$$

4.2. PHENOMENA IN THE SOLAR PHOTOSPHERE

An elementary solution yields for the intensity of radiation emerging in direction μ:

$$I_\nu(\mu) = \int_0^\infty B_\nu(T) e^{-\frac{\tau_\nu}{\mu}} \frac{d\tau_\nu}{\mu} \qquad (4.5)$$

The Planck function must increase with depth, since the temperature increases with depth (se Fig. 4.1). Eddington made the following ansatz:

$$B_\nu = C + D\tau\nu \qquad (4.6)$$

If we put this into 4.5, we arrive at

$$I_\nu = C + D\mu \qquad (4.7)$$

The physical depth z corresponding to $\tau_\nu = 1$ is said to be the origin of the emergent radiation of frequency ν. Thus, by observing the photosphere at different frequencies, we sample it at different heights. The deepest penetration is obtained at IR wavelengths (about 1.6 μm); higher layers may be sampled by observing at the centers of absorption lines. Thus, the surface of the Sun is defined as $\tau_{5000 \text{ Å}} = 1$.

If we look at the solar disk we immediately see that the central regions are brighter than the limb. The function $I_\nu(\mu)/I_\nu(1)$ is called the *limb darkening* (center to limb variation). This may be written as:

$$\frac{I_\nu(\mu)}{I_\nu(1)} = \int_0^\infty \frac{B_\nu(T)}{I_\nu(1)} e^{-\tau_\nu/\mu} \frac{d\tau_\nu}{\mu} \qquad (4.8)$$

If one does an inversion of this equation information about the physical structure (temperature distribution) of the solar atmosphere is obtained. Stellar limb functions can not be measured accurately so this method is only applicable to the Sun.

4.2.2 Granulation

Under very good seeing conditions the Sun shows a cellular like pattern which is called granulation. The mean diameter of the cells is about 1000 km which corresponds roughly to 1 arcsec (as seen from the Earth). In the bright granules matter is streaming upwards, in the darker intergranular lanes streaming downwards. Up to now the best granulation images have been taken from the ground since no large solar telescopes have been launched. In order to minimize the effect of the turbulence of the Earth's atmosphere (seeing), the exposure times must be shorter than 1/10 s. Usually, one makes a burst of several images and then selects the best image for further analysis. Spectrograms show a high degree of correlation between intensities and velocities proving the convective character of the phenomenon. Under a spatial resolution better than 0.5 arcsec, the situation becomes more complex. Regular granules seem to have a maximum for the upflow near their center, so called exploding granules have a maximum upflow between the center and the edge. Measuring the width of spectral lines one gets a hint for turbulence. Enhanced line widths indicate enhanced turbulence. It was found that turbulence is located in the downdrafts which is also predicted by 3 D models. The turbulence may be generated by the shear between upflows and downflows at granular borders and on transonic flows.

A recent review about solar granulation was given by Muller (1999) where further references can be found. A problem to investigate the granulation is how can we identify a granulum? One possibility is to identify them by an isophote contour at a level close to the average intensity of the photosphere. The images must be filtered in order to remove the intensity fluctuations at low frequency, originating in instrumental brightness inhomogeneities and in solar large scale fluctuations (which arise from the supergranulation, mesogranulation and oscillations). Finally, high frequency noise must be eliminated. In the Fourier domain such a filter has the form:

$$F(k) = (1 - e^{-Ca_1^2 k^2}).e^{Ca_2^2 k^2} \tag{4.9}$$

The parameters are chosen, so that the maximum filter transmission stays in between spatial scales 0.5 and 1 arcsec. Such a filter is partially restoring as it enhances the contrast of the smallest granules which can then be identified more clearly. Another method is to find the inflection points of the intensity distribution in the image using a Laplacian operator.

How do granules evolve? The most common process is that of fragmentation: a granule grows and then splits into several fragments (3...4). About 60% of granules appear or die by this process. Some granules appear spontaneously in intergranular spaces and grow, others result from merging of two adjacent granules. The most spectacular evolution is observed for exploding granules.

From the physical point of view, there exists a limitation for the horizontal expansion because of mass conservation and radiative loss. Matter is streaming upward in a granulum, expands and horizontal flows are driven by pressure gradients; thus the central upflow is decelerated which then cannot supply the horizontal expansion and the radiative loss. The central part cools and the granule splits into several fragments, after a downdraft developed. On the other hand, intergranular lanes are interconnected without interruption. They contain some dark holes which exist over 45 min and may correspond to the fingers of downflowing material predicted by 3 D models.

Using time series with the 50 cm refractor at the turret dome of the Pic du Midi observatory Roudier et al. (1997) showed the existence of singularities in the intergranular lanes what they called intergranular holes which have diameters between 0.24 arcsec and 0.45 arcsec and are visible fpr more than 45 min. These holes appear to be systematically distributed at the periphery of mesogranular and supergranular cells. Spectroscopic observations of the solar granulation with high resolution yield information about velocities e.g. when observed near solar disk center, granular profiles are blueshifted because matter rises and moves in direction to the observer (see Fig. 4.2.2).

The granule lifetime can be determined by their visual identification on successive images or by cross correlating these images. There is a large discrepancy of the results: granular lifetimes range from 6 to 16 min.

4.2. PHENOMENA IN THE SOLAR PHOTOSPHERE

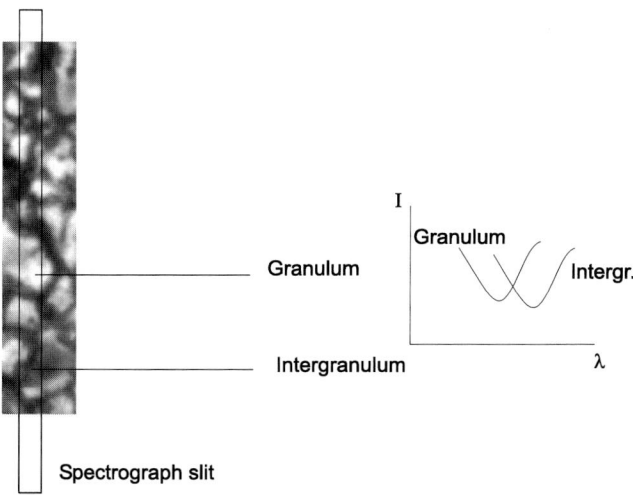

Figure 4.2: Spectroscopic observation of solar granulation. The entrance of a spectrograph slit covers different granular/intergranular areas. Line profiles emanating from granules are blueshifted because matter moves upwards and profiles from intergranular areas are redshifted because matter moves away from the observer. This is valid for solar granulation observed near the disk center.

Concerning the structural properties of granules, we have to mention that their number N increases monotonically with decreasing size. Granules of size 1.4 arcsec are the main contributors to the total granule area. When the area A is plotted versus their perimeter in a log-log scale, the dispersion of points (each of them marks a granule) is small and their shape can be characterized by the relation:

$$P \sim A^{D/2} \quad (4.10)$$

where D is the fractal dimension. It seems that there are two ranges with different fractal dimensions:

- $D \sim 1.25$ for granules smaller than about 1.35 arcsec.

- $D \sim 2.00$ for granules that are larger than 1.35 arcsec.

The physical interpretation is as follows: In hydrodynamics, the fractal dimension is often used to get some information about the dynamical state. In the theory of Kolmogorov (he treated isotropic, homogeneous turbulence in three dimensions and obtained a 5/3 power law for the energy spectrum) a value of $D = 5/3$ is predicted for isotherms and 4/3 for isobars. A fractal dimension of 2 or even larger means that the shape is complex which is confirmed by observations since many of them are in the process of fragmentation.

Granules above 1.4 arcsec have nearly the same brightness, the intergranular brightness is nearly constant, with an average value of 0.92 (when the averaged continuum is at 1.0). The rms intensity fluctuations of the best image is 10-11% at $\lambda\, 4650\,\text{Å}$ (50 cm refractor at La Palma) and 8-9% at $\lambda\, 5700\,\text{Å}$ (50 cm refractor at the Pic du Midi). Restored values lie between 10 and 22%. From the granular contrast we can infer the temperature variations (assuming Planck's law) which correspond to $\sim 200\,\text{K}$.

Theoretical approaches

The simplest model of convection is the classical Rayleigh problem: suppose that we have a fluid (either gaseous or liquid), confined between two horizontal plates separated by a distance h and maintained at temperature T_1 (upper) and T_2 (lower) with $T_2 > T_1$. If the fluid has a positive coefficient of thermal expansion α as it will be the case for a gas and for a normal fluid, the fluid near the lower plate will tend to rise. However, this will be opposed by two effects: a) viscous dissipation, b) thermal diffusion in the fluid. Convection will occur when the imposed temperature gradient $(T_2 - T_1)/h$ is sufficiently large or, for a given gradient, when the coefficients of the kinematic viscosity ν and of thermal diffusion κ are sufficiently small. Rayleigh's theoretical analysis of the problem in 1916 inspired Bénard to investigate this 40 years later. It was found that convective instability occurs when the Rayleigh number R exceeds a critical value:

$$R > R_{\text{crit}} \qquad R = \frac{g\alpha\beta h^4}{\kappa\nu} \qquad (4.11)$$

where β is the temperature gradient. For R_{crit} Rayleigh found the value 657.5. This value depends on the boundary conditions. Later Chandrasekhar has shown that e.g. a Coriolis force (as an effect of rotation) inhibits the onset of instability to an extend which depends on the value of a non dimensional parameter (called Taylor number):

$$C = \frac{4h^4\Omega^2}{\nu^2} \qquad (4.12)$$

here, Ω is the vertical component of the angular velocity vector. For details see e.g. Chandrasekhar (1961).

Let us discuss the Rayleigh number for the solar convection zone. The value is extremely high, $R \sim 10^{10\ldots 11}$.

Important information about the origin of the solar granulation can be inferred from power spectra. From spectrograms we can obtain 1-D power spectra of intensity and velocity fluctuations, from white light images, one gets 2-D power spectra for the intensity fluctuations. The theoretical power spectrum of the velocity fluctuations decreases as $k^{-5/3}$ down to the scale of molecular diffusion. The temperature power spectrum however decreases as $k^{-5/3}$ only to a scale k_c. At smaller scales the spectrum decreases as $k^{-17/3}$. Thus k_c separates the inertial convective range, where heat advection dominates from the inertial conductive range, where diffusion dominates. The former is the range of large granules, the latter the range of small granules.

The basic set of hydrodynamic equations to describe solar convection is as follows:

4.2. PHENOMENA IN THE SOLAR PHOTOSPHERE

- Conservation of mass

$$\frac{\partial \ln \rho}{\partial t} = -\vec{v}.\nabla \ln \rho - \nabla.\vec{v} \qquad (4.13)$$

- Conservation of momentum:

$$\frac{\partial \vec{v}}{\partial t} = -\vec{v}.\nabla \vec{v} + \vec{g} - \frac{P}{\rho}\nabla \ln P + \nabla.\sigma \qquad (4.14)$$

- Conservation of energy:

$$\frac{\partial e}{\partial t} = -\vec{v}.\nabla e - \frac{P}{\rho}\nabla.\vec{v} + Q_{\rm rad} + Q_{\rm visc} \qquad (4.15)$$

$Q_{\rm rad}$ is obtained from the equation of transfer and is the ratio of radiative heating/cooling.

The ionization energy dominates the internal energy e near the surface (Stein and Nordlund, 2000).

Interaction between granulation and magnetic elements

In this section we consider magnetic regions which occur as Plages or faculae (in active regions) and in the photospheric network (in the quiet Sun) in the form of small bright points. Sunspots will be discussed in the next paragraph. Magnetic elements (observed in high resolution magnetograms) and bright points (observed in high resolution filtergrams) coincide. Bright points are visible in white light near the limb (e.g. as faculae) but also at the disk center because they have a brightness comparable to granules. It is very easy to observe them with a G Band filter.

The dynamics of the granules forces these small bright points to appear and stay in the intergranulum when the surrounding granules converge. Thus there seems to be a continuous interaction between granules and magnetic elements. Small magnetic flux tubes are the channels along which the energy is carried in upper layers by different kinds of waves. In that context Choudhuri et al. (1993) discussed the generation of magnetic kink waves by rapid footpoint motions of the magnetic flux tubes. They found that these pulses are most efficient. Kalkofen (1997) discussed the impulsive generation of transverse magneto-acoustic waves in the photosphere, propagating upward with exponential growth of amplitude.

Granulation-Mesogranulation

Idealized numerical experiments on turbulent convection were made by Cattaneo et al. (2001). The authors found two distinct cellular patterns at the surface. Energy-transporting convection cells (corresponding to granules in the solar photosphere) have diameters comparable to the layer depth, while macrocells (corresponding to mesogranules) are several times larger. The motion acts as a small-scale turbulent

32 CHAPTER 4. PHENOMENOLOGY OF SOLAR ACTIVE REGIONS

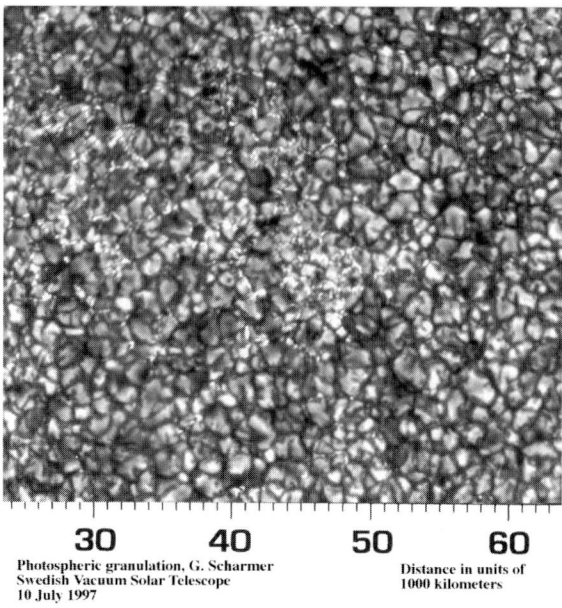

Figure 4.3: Solar granulation and small network bright points

dynamo, generating a disordered magnetic field that is concentrated at macrocellular corners and, to a lesser extent, in the lanes that join them. These results imply that mesogranules owe their origin to collective interactions between the granules.

4.2.3 Five Minutes Oscillations

In 1962 Leighton, Noyes and Simon identified a strong oscillatory component which they called five minutes oscillations because of its characteristic period. Later, these were interpreted as standing acoustic waves trapped in resonant cavities below the photosphere. This is the field of helioseismology that will be discussed in a different chapter.

The spatial relation between the 5-min oscillations and the granulation pattern has been largely debated in the literature. Of course such a discussion is important to understand the excitation mechanism of these oscillations and, hence, the internal properties of the Sun. Theoretical studies suggest that acoustic waves which comprise the 5-min oscillations are stochastically generated by turbulent convection just beneath the photosphere (Goldreich et al, 1994). Espagnet et al. (1996) studied the relation between oscillation and granulation and found that the most energetic oscillations are concentrated in downflow regions in expanding intergranular spaces. This was later confirmed by Goode et al. (1998).

Strous et al. (2000) found a roughly linear relation between the peak seismic flux and the peak downward convective velocity associated with each seismic event.

Other authors like e.g. Hoekzema which analyzed G band images found that photospheric 5 min oscillations are global and rather insensitive to local fine structure (1998).

A recent paper on that topic was given by Khomenko et al.(2001) where further references can be found.

4.2.4 Sunspots

Discovery of sunspots

When the Sun is very low just above the horizon one can make a short glimpse on it with the unprotected naked eye. Chinese astronomers were the first who reported on dark spots visible on the Sun. In the year 1611 sunspots were observed for the first time through a telescope and four men can be named as the discoverers: J. Goldsmid (Holland), G. Galilei (Italy), Ch. Scheiner (Germany) and Th. Harriot (England). The first publication on that topic appeared from Goldsmid (he is better known by his Latin name Fabricius). He even argued that the Sun must rotate since the sunspots move across the disk. Since he was a Jesuit he first suspected some defect in his telescope when he observed the spots. Then he failed to persuade his ecclesiastical superiors who refused to allow him to publish his discovery. However, Scheiner announced his discovery in three anonymous letters to a friend of Galileo and Galileo responded in three letters in 1612 (the sunspot letters) that he had discovered the sunspots. Of course Scheiner and Galileo became enemies. Scheiner later reported his discoveries in his work Rosa Ursinae sive Sol in 1630. Both scientists noted that the spots appear only within zones of low latitudes at either side of the equator. There are never spots near the poles.

Of course one never should risk a blinding of the eyes and the safest way to observe the Sun with a telescope is when the solar image at the ocular (exit pupil) is projected onto a screen.

After the initial interest and the publication of Scheiner's major work there was no big interest in sunspots. In 1977 Eddy showed that this must be seen in connection with the fact that during 1640-1705 there was a great reduction in the number of sunspots seen on the Sun which is now know as the *Maunder Minimum*.

The next significant discovery was made by Schwabe who was a German apothecary and bought a telescope in 1826 in order to search for a planet inside the orbit of Mercury. He recorded the occurrence of sunspots over 43 years and reported on a periodicity of their occurrence of about 10 years. In 1851 appeared his publication on the 11 year periodicity of the annually averaged sunspot numbers. Several years later Carrington showed from his observations that the Sun rotates differentially; a point at the equator rotates more rapidly than one at higher latitudes. He defined an arbitrary reference point on latitude 10^0 as longitude zero and a rotation completed by this point is known as Carrington rotation (CR). He was also the first to see a white light flare on the Sun in Sep. 1859 during sketching sunspot projections with a friend. Suddenly two crescent-shaped patches broke out, brightened, moved a distance twice their length, then faded away as two dots. All that happened within a period over five minutes. Carrington reported to the Royal Astronomical Society that at 4 hours after midnight the magnetic instruments indicated a great magnetic storm. So he was in fact the first who noticed that there exists a connection between solar phenomena and disturbances on Earth.

R. Wolf of Bern (1816-1893) studied all available records and derived a more accurate estimate for the sunspot cycle. In 1848 he introduced the relative (Zurich) sunspot number R_Z as a measure for solar activity. Sunspot often appear as groups. If g denotes the number of sunspot groups and f the number of individual spots, then

$$R_Z = k(10g + f) \qquad (4.16)$$

k... personal reduction factor. Today more than 30 observatories contribute to determine this value. Further details about the history of sunspots can be found in the following chapters.

The physics of sunspots

Sunspots consist of dark central regions, called umbra and a surrounding less dark filamentary region called penumbra. The umbral diameter is about 10 000 km but for the largest spots may exceed 20 000 km. Penumbral diameters are in the range of 10 000 -15 000 km. First it was thought that they are something like clouds in front of the solar surface. They drift across the surface of the Sun and this motion can easily be explained by assuming that they are fixed to the sun and drift because the Sun rotates. Sunspots evolve and some of them are visible over more than 1 rotation period. The observations of sunspots showed that the rotation of the Sun is not like that of a solid body. Near the equator it rotates faster (27.7 days) and at a latitude of 40^0 the rotation period is 28.6 days; this is called the differential solar rotation.

Another interesting phenomenon is the *Wilson depression*. Wilson observed (1769) a very large spot nearing the west limb and noted that the penumbra on the further side from the limb gradually contracted and finally disappeared. When the spot reappeared at the east limb some two weeks later, the same behavior was displayed by the penumbra on the opposite site of the spot. The surface of a sunspot is depressed below the surface of the surrounding plasma.

The temperature of the umbra is about 4 000 K whereas the temperature of the solar surface is about 6 000 K. According to Stefan's law the total energy emitted per unit area by a black body at temperature T is proportional to T^4; the above mentioned temperature difference between umbra and photosphere means that the energy flux through a given area of the umbra is $\sim 20\%$ of that through an equivalent area of the photosphere. The penumbra has a temperature between umbra and solar surface. In the penumbra we observe also the *Evershed effect* which means a radial outflow of matter with the velocity increasing outwards with a characteristic speed of 1 to 2 kms/s.

In 1908 Hale discovered that the spectral lines were split in the sunspots. This is caused by the *Zeeman effect* in the presence of strong magnetic fields. In the absence of magnetic fields several quantum mechanical state may possess the same

4.2. PHENOMENA IN THE SOLAR PHOTOSPHERE

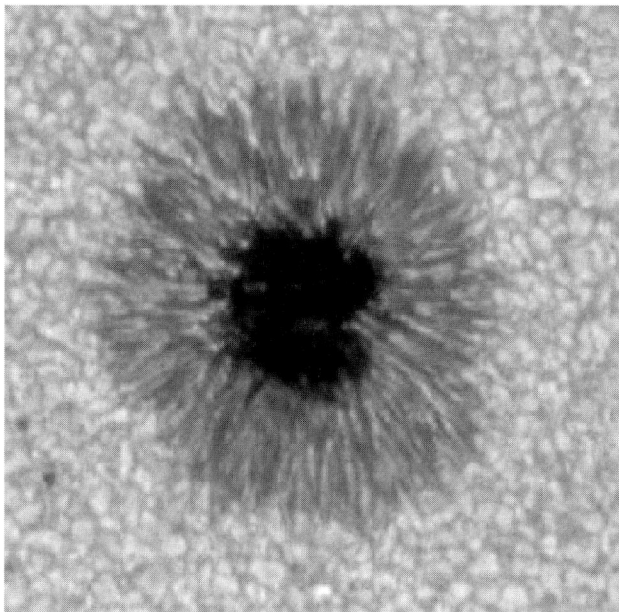

Figure 4.4: Large sunspot showing the dark central umbra and the filamentary penumbra. Outside the penumbra the granulation pattern is clearly seen

energy but the magnetic fields destroys this symmetry and one has a splitting of the energy levels. The displacement of the lines due to the Zeeman effect is given by:

$$\Delta\lambda = 4.7 \times 10^{-8} g^* \lambda^2 B \qquad (4.17)$$

The wavelength λ is given in nm, the Landéfactor g^* depends on the spin and orbital momentum of the levels and B denotes the magnetic induction given in Tesla.

The strength of the magnetic field is in the order of 3 000 Gauss.

Small dark spots with diameters < 2 500 km lacking penumbrae are called pores. They exist within groups or appear also as isolated structures. Their lifetimes are in the range of a few hours to several days.

Sunspot groups tend to emerge either sequentially at the same or similar Carrington longitudes, which are designated as active longitudes, or to overlap in clusters.

High spatial observations of spots

High spatial resolution observations of sunspots show that there appear a lot of different morphological phenomena: multiple umbrae, bright umbral dots, light bridges, dark nuclei in the umbra etc. One problem in the study of sunspots and their fine structure is observational stray light.

An important photometric parameter of umbral cores is the minimum intensity (intensity of the darkest point) $I_{\min}$ which is usually in the range of 0.05-0.3 of the mean photospheric intensity at $\lambda \sim 5\,400$ Å. There seems to be a relation between the size of the umbrae and the temperature. Umbrae with a diameter $D_U < 7''$ have higher temperatures than the large ones. Moreover, regions with higher magnetic field strength are darker and cooler than those with lower strength. The darkest regions in umbral cores are dark nuclei. These are the areas with the strongest magnetic fields and the orientation of the field is perpendicular to the surface of the Sun. They are not necessarily centered in the umbral cores, some of them are observed close to the edge of the penumbra. They cover 10-20% of the total umbral core area and their size is about 1.5 arcsec.

The penumbra shows elongated structures which is a consequence of the strongly inclined magnetic field. Bright penumbral filaments consist of penumbral grains. They seem to have cometary like shapes with "heads" pointing towards the umbra and have a mean width of only $0.36''$ and a length of $0.5...2''$. The observed brightness approaches the photospheric one and the lifetimes are between 40 minutes and 4 hours. They are separated by narrow dark fibrils. The magnetic field seems to be stronger and more horizontal in dark fibrils and weaker and more vertical in penumbral grains.

It is also interesting to note that nearly all penumbral fine structures are in motion. The penumbral grains move towards the umbra with an average speed of 0.3-0.5 km/s. On the other hand, dark cloud like features which arise from the dark fibrils move rapidly outwards (up to 3.5 km/s) towards the outer penumbral border.

The last fine structure which is important to study are the light bridges. They cross the umbra or penetrate deeply into it and can be observed for several days although they change their shape substantially on the scale of hours. They can be classified into faint (located inside umbral cores) and strong (separating umbral cores). Strong light bridges separate umbral cores of equal magnetic polarities and a subclass of them opposite polarities. The analysis of 2-D power spectra of intensity fluctuations inside strong light bridges showed that the "granules" that can be seen there are smaller (1.2 arcsec, normal granulation: 1.5 arcsec) and the slopes of power spectra indicated the presence of a Kolmogorov turbulent cascade. The magnetic field strength in strong light bridges is substantially lower than in adjacent umbra.

A recent review about the fine structure of sunspots was given by Sobotka (1999) where other references can be found. A review on empirical modelling and thermal structure of sunspots was given by Solanki (1997).

Sunspots and magnetic fields

Observations demonstrated, that spots often occur in bipolar magnetic groups. The magnetic polarity of the leading spot in the pairs (in terms of solar rotation) changes from one 11 year cycle to the next- this is know as *Hale's law*. There is a 22 year magnetic cycle. Spots appear as a magnetic flux tube rises (see magnetic buoyancy) and intersects with the photosphere. The magnitude of the magnetic induction is 0.3 T in the umbra and 0.15 T in the penumbra. In the umbra the field is approximately vertical, and the inclination increases through the penumbra.

Hale's observations also suggested that the Sun has an overall dipolar magnetic field (10^{-4} T). This very weak dipolar field is reversed over the magnetic cycle. Almost all of the photospheric field outside sunspots is concentrated in small magnetic elements with a magnetic induction between 0.1 and 0.15 T.

Only the surface properties of the flux tube that defines a spot can be observed. The question is, how the field structure changes with depth. The simplest model is a monolithic column of flux. Let us assume that the pressure inside the flux tube is negligible compared to the magnetic pressure. We also assume that the gravitational force is unimportant in obtaining an approximate idea of the magnetic field structure, the magnetic field in cylindrical polar coordinates can be taken to be current free:

$$\vec{B} = \frac{1}{\omega}\left[-\frac{d\psi}{dz}, 0, \frac{\partial \psi}{\partial \omega}\right] \quad (4.18)$$

Thus curl$\vec{B} = \vec{0}$. Since div$\vec{B} = 0$,

$$\frac{\partial^2 \psi}{\partial \omega} - \frac{1}{\omega}\frac{\partial \psi}{\partial \omega} + \frac{\partial^2 \psi}{\partial z^2} = 0 \quad (4.19)$$

The neighboring photosphere, in which the flux tube is embedded has a known pressure variation with height $P_e(z)$. The boundary of the flux tube is at $\omega = \omega_0(z)$, where

$$B^2/2\mu_0 = P_e(z) \quad (4.20)$$

We see that as $z \to \infty$ the field becomes nearly horizontal and $B_\omega \sim F/2\pi\omega_0^2$ and as $z \to -\infty$, the field becomes vertical and $B_\omega \sim F/\pi\omega_0^2$.

There is one problem with this monolithic model: the difference in the energy radiated by the spot and by an equivalent area of the normal photosphere is only about a factor of 4. This is less than would be expected if convection in the spot were completely suppressed. Therefore, it is believed that some form of convective energy transport must occur and the field must be more complex e.g. coherent flux tube or a tight cluster. Reviews about these topics were given by Bogdan (2000) and Hurlburt (1999).

Sunspot group classification

The 3 component McIntosh classification (McIntosh, 1990) is based on the general form 'Zpc', where 'Z' is the modified Zurich Class, 'p' describes the penumbra of the principal spot, and 'c' describes the distribution of spots in the interior of the group. This classification scheme substituted the older scheme that was introduced by Waldmeier.

1. Z-values: (Modified Zurich Sunspot Classification).

    - A - A small single unipolar sunspot. Representing either the formative or final stage of evolution.
    - B - Bipolar sunspot group with no penumbra on any of the spots.

- C - A bipolar sunspot group. One sunspot must have penumbra.
- D - A bipolar sunspot group with penumbra on both ends of the group. Longitudinal extent does not exceed 10 deg.
- E - A bipolar sunspot group with penumbra on both ends. Longitudinal extent exceeds 10 deg. but not 15 deg.
- F - An elongated bipolar sunspot group with penumbra on both ends. Longitudinal extent of penumbra exceeds 15 deg.
- H - A unipolar sunspot group with penumbra.

2. p-values:

- x - no penumbra (group class is A or B)
- r - rudimentary penumbra partially surrounds the largest spot. This penumbra is incomplete, granular rather than filamentary, brighter than mature penumbra, and extends as little as 3 arcsec from the spot umbra. Rudimentary penumbra may be either in a stage of formation or dissolution.
- s - small, symmetric (like Zurich class J). Largest spot has mature, dark, filamentary penumbra of circular or elliptical shape with little irregularity to the border. The north-south diameter across the penumbra is less or equal than 2.5 degrees.
- a - small, asymmetric. Penumbra of the largest spot is irregular in outline and the multiple umbra within it are separated. The north-south diameter across the penumbra is less or equal than 2.5 degrees.
- h - large, symmetric (like Zurich class H). Same structure as type 's', but north-south diameter of penumbra is more than 2.5 degrees. Area, therefore, must be larger or equal than 250 millionths solar hemisphere.
- k - large, asymmetric. Same structure as type 'a', but north-south diameter of penumbra is more than 2.5 degrees. Area, therefore, must be larger or equal than 250 millionths solar hemisphere.

3. c-values:

- x - undefined for unipolar groups (class A and H)
- o - open. Few, if any, spots between leader and follower. Interior spots of very small size. Class E and F groups of 'open' category are equivalent to Zurich class G.
- i - intermediate. Numerous spots lie between the leading and following portions of the group, but none of them possesses mature penumbra.
- c - compact. The area between the leading and the following ends of the spot group is populated with many strong spots, with at least one interior spot possessing mature penumbra. The extreme case of compact distribution has the entire spot group enveloped in one continuous penumbral area.

4.2. PHENOMENA IN THE SOLAR PHOTOSPHERE

There exists also the Mount Wilson classification scheme:

- α: Denotes a unipolar sunspot group.

- β: A sunspot group having both positive and negative magnetic polarities, with a simple and distinct division between the polarities.

- $\beta - \gamma$: A sunspot group that is bipolar but in which no continuous line can be drawn separating spots of opposite polarities.

- δ: A complex magnetic configuration of a solar sunspot group consisting of opposite polarity umbrae within the same penumbra.

- γ: A complex active region in which the positive and negative polarities are so irregularly distributed as to prevent classification as a bipolar group.

Sunspots and the Solar Cycle

The number of sunspots changes with a 11 years period. Today we know that all solar activity phenomena are related to sunspots and thus to magnetic activity. To measure the solar activity the sunspot numbers were introduced:

$$R = k(10g + f) \tag{4.21}$$

Here g denotes the number of sunspot groups and f the number of spots. The factor k is a correction which takes into account for the different instruments used for the determination of R. In order to smear out effects of solar rotation, R is given as a monthly averaged number and called the sunspot relative number. Today there exist better methods to quantify the solar activity however sunspot numbers are available for nearly 400 years and thus this number is still used.

The Royal Greenwich Observatory (RGO) compiled sunspot observations from a small network of observatories to produce a data set of daily observations starting in May of 1874. The observatory concluded this data set in 1976 after the US Air Force (USAF) started compiling data from its own Solar Optical Observing Network (SOON). This work was continued with the help of the US National Oceanic and Atmospheric Administration (NOAA) with much of the same information being compiled through to the present.

Since 1981, the Royal Observatory of Belgium harbors the Sunspot Index Data center (SIDC), the World data center for the Sunspot Index. Recently, the Space Weather forecast center of Paris-Meudon was transferred and added to the activities of the SIDC. Moreover, a complete archive of all images of the SOHO instrument EIT has become available at the SIDC.

Let us briefly summarize the behavior of sunspots during the activity cycle:

- The leader spots (i.e. by convention it is defined that the Sun rotates from east to west; the largest spot of a group tends to be found on the western side and is called the leader, while the second largest in a group is called the follower) in each hemisphere are generally all of one polarity, while the follower spots are of the opposite polarity.

CHAPTER 4. PHENOMENOLOGY OF SOLAR ACTIVE REGIONS

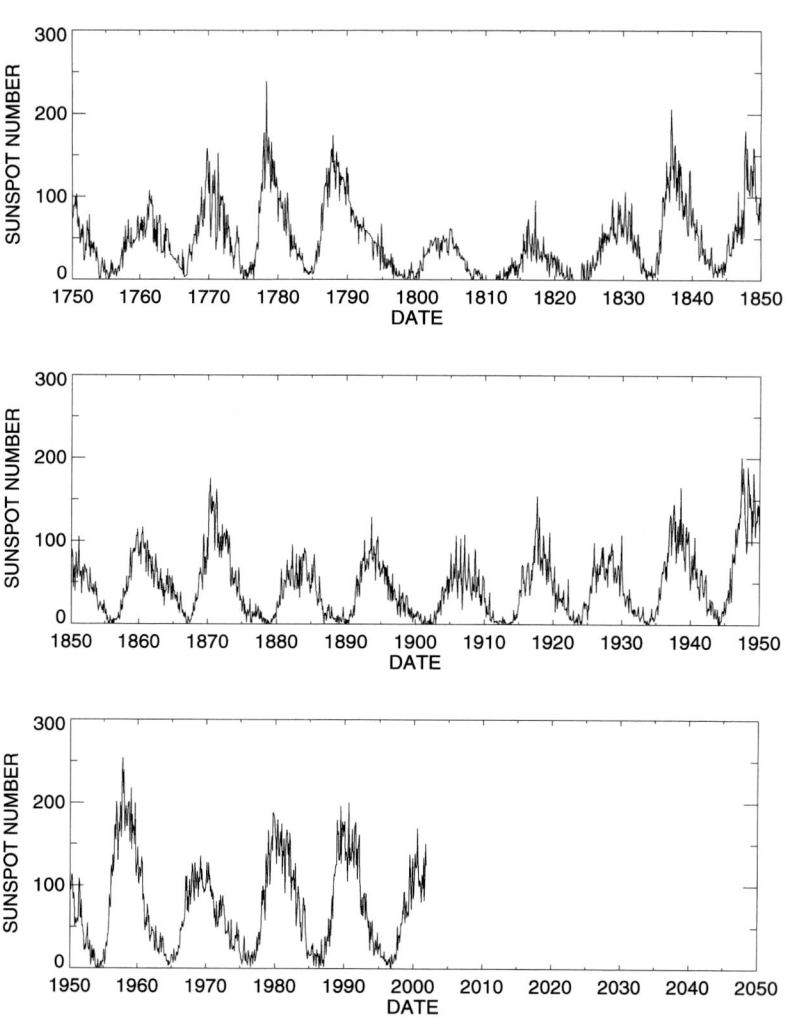

Figure 4.5: Relative Sunspot number

4.2. PHENOMENA IN THE SOLAR PHOTOSPHERE

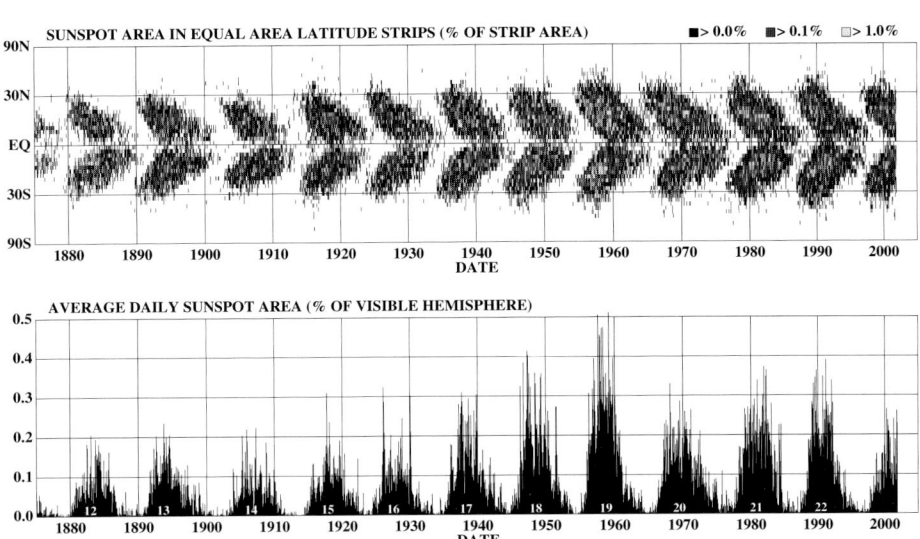

Figure 4.6: Butterflydiagram illustrating the equatorward motion of spots during the activity cycle.

- If the leaders and followers are regarded as magnetic bipoles, the orientation of these bipoles is opposite on opposite hemispheres.

- The magnetic axes of the bipoles are inclined slightly towards the equator, the leader spot being closer. This inclination is about 12^0.

- Towards the end of a cycle spot groups appear at high latitudes with reversed polarity, they belong to the new cycle whereas those with normal polarity for the old cycle occur close to the equator. This is illustrated in the so called butterfly diagram (see Fig. 4.6).

4.2.5 Photospheric Faculae

Near the solar limb, regions brighter than the surrounding photosphere can be found and are known as photospheric faculae. These structures are hotter than their surroundings. At the disk center they are not visible. In the neighborhood of sunspots they tend to overlap and can be identified further from the limb. They appear in increased numbers in a region prior to the emergence of sunspots and remain for a rotation or more after the spots have decayed. They are important for the energy balance between sunspots and the photosphere. Faculae can be observed on the whole disk using filtergrams. In that case they are often called plage and attributed to the chromosphere. Photospheric faculae are manifestations of concentrated azimuthal magnetic fields. One possibility to study sunspots and faculae at photospheric levels is to use the Ca II K line 0.5 Å off the center with a 0.15 Å passband.

Polar faculae appear as pointlike, bright photospheric spots near the solar limb at latitudes of 55 degrees or more (average of 65 degrees). Polar faculae tend to occur at lower latitudes (as low as 45 degrees) during the years in which there are only few observable. They can be distinguished from main zone faculae by their essentially pointlike and solitary appearance, in contrast to the more area- and grouplike appearance of the main zone faculae (55 degrees or lower). They have a shorter lifetime (minutes to hours) than ordinary faculae. The brightest can last for a couple of days, and can be traced farther from the solar limb too. In connection with the activity cycle it is interesting to note that polar faculae are most numerous at times of minimum solar activity, which in turn might be an additional hint for their relation with the upcoming new solar cycle.

4.3 The Chromosphere

The chromosphere can be observed during short phases of solar eclipses. The spectrum obtained at these rare occasions is called a flash spectrum. The chromosphere lies between the corona and the photosphere. Above the photosphere the temperature passes through a minimum of $4\,000$ K and then rises to many 10^4 K in the chromosphere and much more rapidly in the transition region until the coronal temperature ($\sim 10^6$ K) is reached. Two very prominent spectral lines formed in the chromosphere are the so called H and K lines of singly ionized Ca (called Ca II). These lines are in absorption in the spectrum of the photosphere but appear as emission lines in the hotter chromosphere. Their strength varies through the sunspot cycle, the lines are stronger at maximum. The observations of the variation of the strength of stellar H and K lines provide thus information about stellar activity cycles. Important chromospheric lines are given in Table 4.1, the physics of the formation of these lines is complicated since the assumption of LTE is not valid.

The temperature variation throughout the chromosphere can be described as follows:

- Temperature minimum: near 500 km; here the UV continuum near 1 600 Å, the far IR continuum and the minima in the wings of Ca II and Mg II lines are formed,

- moderately fast temperature increase from T_{min} to approx. 6 000 K. In the first plateau there are the emission peaks of Ca II and Mg II, the centre of $H\alpha$, the mm continuum and the wing of Lyα.

- temperature plateau near $6\,000$ - $7\,000$ K

- sharp temperature rise beginning near $8\,000$ K and terminating in a thin plateau near $22\,000$K. From the second plateau the central portion of Lyα and the 3 cm continuum is emitted

Table 4.1: Prominent chromospheric emission lines

Line	Wavelength
Lyα	1216 Å
Lyβ	1026 Å
C I continua	≤ 1100 Å, ≤ 1239 Å
Mg II h	2803 Å
Mg II k	2796 Å
Ca II H	3968 Å
Ca II K	3934 Å
He I	4471 Å , 5876 Å
Ca II IR	8498, 8542, 8662 Å
Mg I b$_{1,23}$	b$_2$ 5173 Å
Na D$_{1,2}$	5896, 5890 Å
Hα	6563 Å
Hβ	4861 Å
CO	4.6μ

Thus by observing in different lines or even in different depths of a particular line, one can probe the chromosphere at different height levels. As it is indicated above, it is possible to observe the chromosphere in radio waves at mm to cm wavelengths. The emission processes here are free free transitions of electrons with a Maxwellian distribution.

When analyzing the H and K lines bright grains are detected. These bright grains are produced by shocks near 1 Mm (10^6 m) height in the chromosphere.

4.3.1 Radiative Transfer in the Chromosphere

Above the temperature minimum, the spectral lines are formed under non local thermodynamic equilibrium conditions (NLTE).

Let us start with the change of the specific intensity I_ν along a short distance ds: there will occur absorption and emission, both of which are described by the coefficients:

- κ_ν absorption coefficient

- η_ν emission coefficient

For simplicity we consider a homogeneous, plane-parallel atmosphere stratified by gravity. Then, the properties depend only on the height z. The surface of the atmosphere in a strict mathematical sense is where no interactions take place, i.e. the particle densities are extremely low. The optical depth is defined by:

$$d\tau_\nu = -\kappa_\nu dz, \qquad \tau_\nu = -\int_\infty^z \kappa_\nu dz' \qquad (4.22)$$

The source function is the ratio between the two coefficients:

$$S_\nu = \eta_\nu / \kappa_\nu \qquad (4.23)$$

In local thermodynamic equilibrium (LTE) we have the relation:

$$S_\nu = B_\nu(T) \qquad (4.24)$$

which is called Kirchhoff's law, $B_\nu(T)$ being the Planck function. We can progress to solve the transport equation:

$$I_\nu(\tau_\nu = 0, \mu) = \int_0^\infty S_\nu(\tau'_\nu) e^{-\tau'_\nu/\mu} d\tau'_\nu / \mu \qquad (4.25)$$

In this equation $\mu = \cos\theta$, θ being the angle between the normal to the disk center and the point where observations are done.

From a Taylor series expansion of S_ν about a not specified τ^*_ν one gets

$$I_\nu \sim S_\nu(\tau_\nu) = \mu \qquad (4.26)$$

where τ^*_ν was specified to μ. That means, one observes under the angle θ to z approximately the source function at optical depth $\tau_\nu = \mu$.

Let us consider two energy levels in an atom which have the quantum numbers l (lower level) and u (upper level). The number of atoms per cm^3 in the lower level is N_l und in the upper level N_u. Of course a transition from l to u corresponds to an absorption process, where a photon of energy $h\nu_{l,u} = \chi_u - \chi_l$ is absorbed. Thus the number of transition per cm^3 is given by:

$$n(l \to u) = N_l J_{\nu(l,u)} B(l,u) \qquad (4.27)$$

$B(l,u)$ is the transition probability for the transition $l \to u$. On the other hand let us consider the number of spontaneous transitions from $u \to l$ which is independent on the intensity J:

$$n(u \to l) = N_u A(u,l) \qquad (4.28)$$

$A(u,l)$ is the transition probability for spontaneous transitions. Generally, we do not know what the average intensity $J_{\nu(l,u)}$ is, only in thermodynamic equilibrium it is equal to the Planck function. In thermodynamic equilibrium there is a direct balancing between the number of transitions $u \to l$ and $l \to u$ and the ratio of the occupation numbers is governed by the Boltzmann formula:

$$\frac{N_u}{N_l} = \frac{g_u}{g_l} e^{-(\chi_u - \chi_l)/kT} \qquad (4.29)$$

and

$$n(l \to u) = n(u \to l) \qquad (4.30)$$

$$N_l \frac{2h\nu^3}{c^2} \frac{1}{e^{h\nu/kT} - 1} B(l,u) = N_u A(u,l) \qquad (4.31)$$

4.3. THE CHROMOSPHERE

where we have put the Planck function. Let us also substitute the Boltzmann formula:

$$\frac{2h\nu^3}{c^2}\frac{1}{e^{h\nu/kT}-1} = \frac{g_u}{g_l}e^{-(\chi_u-\chi_l)/kT}\frac{A(u,l)}{B(l,u)} \quad (4.32)$$

$$= \frac{g_u}{g_l}e^{-h\nu_{u,l}/kT}\frac{A(u,l)}{B(l,u)} \quad (4.33)$$

where g_u, g_l are the statistical weights of the states u, l. This was first found by Einstein. We must also consider the induced emission which are transitions from $u \to l$ depending on the intensity J. The number of induced emissions is written as:

$$n'(u \to l) = N_u B(u,l) J_{\nu(u,l)} \quad (4.34)$$

In an induced emission process, the photons emitted have the same directions and phases as the inducing photons. Thus a detailed balancing in thermodynamic equilibrium reads as:

$$N_l J_{\nu(u,l)} B(l,u) - N_u J_{\nu(u,l)} B(u,l) = N_u A(u,l) \quad (4.35)$$

and using $J_{\nu(u,l)} = B_\nu$ and the Boltzmann formula:

$$\frac{2h\nu^3}{c^2}\frac{1}{e^{h\nu/kT}-1}\left(B(l,u)\frac{g_l}{g_u}e^{h\nu_{u,l}/kT} - B(u,l)\right) = A(u,l) \quad (4.36)$$

$$B(u,l)g_u = B(l,u)g_l \quad (4.37)$$

$$A(u,l) = B(l,u)\frac{g_l}{g_u}\frac{2h\nu_{u,l}^3}{c^2} = B(u,l)\frac{2h\nu_{u,l}^3}{c^2} \quad (4.38)$$

These relations are called Einstein transition probabilities.

$B(u,l), B(l,u), A(u,l)$ are atomic constants. Though these relations were derived from thermodynamic equilibrium, they must always hold. Therefore, they can be used to get information for excitation conditions and the source function in case we do not have thermodynamic equilibrium.

Let us obtain the source function for a given transition between two energy levels u, l. The coefficient ϵ_ν includes spontaneous transitions as well as induced emission processes:

$$S_\nu = \epsilon_\nu/\kappa_\nu \quad (4.39)$$

Sometimes the induced emission process is included as negative absorption, therefore we then write:

$$S_\nu = \epsilon'_\nu/\kappa'_\nu \quad (4.40)$$

Let us assume that the source function within each transition between two energy levels is independent of frequency, i.e. the source function in a given line is assumed to be frequency independent which means ϵ' and κ' have the same frequency dependence, writing

$$\epsilon'_\nu = \epsilon'_0 \phi(\nu) \qquad \kappa'_\nu = \kappa'_0 \phi(\nu) \quad (4.41)$$

and
$$\int_{\text{line}} \phi(\nu)d\nu = 1 \tag{4.42}$$
this integral has to be extended over all frequencies within the line. Then we find:
$$S_\nu = \epsilon'_0/\kappa'_0 \tag{4.43}$$
The net number of absorbed photons is
$$\begin{aligned}
N(l \to u) - n'(u \to l) &= N_l \bar{J}_\nu B(l,u) - N_u B(u,l) \bar{J}_\nu \\
&= N_l \bar{J}_\nu B(l,u) \left(1 - \frac{N_u}{N_l} \frac{B(u,l)}{B(l,u)}\right) \\
&= N_l \bar{J}_\nu B(l,u) \left(1 - \frac{N_u}{N_l} \frac{g_u}{g_l}\right) \\
&= 4\pi \frac{\bar{J}_\nu}{h\nu} \int_{\text{line}} \kappa'_\nu d\nu \\
&= 4\pi \frac{\bar{J}_\nu}{h\nu} \kappa'_0
\end{aligned}$$

The number of photons spontaneously emitted per cm^3 is
$$\frac{4\pi}{h\nu} \int_{\text{line}} \epsilon'_\nu d\nu = \frac{4\pi}{h\nu} \epsilon'_0 = N_u A(u,l) \tag{4.44}$$

This gives us:
$$\begin{aligned}
S_\nu &= \epsilon'_0/\kappa'_0 = \frac{N_u A(u,l) h\nu}{4\pi} \frac{4\pi}{N_l B(l,u) \left(1 - \frac{N_u g_l}{N_l g_u}\right) h\nu} \\
&= \frac{N_u}{N_l} \frac{A(u,l)}{B(l,u)} \frac{1}{\left(1 - \frac{N_u g_l}{N_l g_u}\right)} \\
&= \frac{N_u}{N_l} \frac{g_l}{g_u} \frac{2h\nu^3}{c^2} \frac{1}{\left(1 - \frac{N_u g_l}{N_l g_u}\right)}
\end{aligned}$$

and after simplifying:
$$S_\nu = \frac{2h\nu^3}{c^2} \frac{1}{\left(\frac{N_l g_u}{N_u g_l} - 1\right)} \tag{4.45}$$

Please note, that in deriving that equation we have not made any assumption of an equilibrium condition. This expression is quite general.

How to determine the occupation numbers N_u/N_l? The number of collisional excitation per cm^3 and s is described by:
$$n_c(l \to u) = N_l \int_0^\infty n_e(v) C'_{l,u}(v) dv = N_l n_e C_{l,u}(T) \tag{4.46}$$

4.3. THE CHROMOSPHERE

Here $C'_{l,u}(v)$ is the probability for an excitation of an electron in level l into a level u by a collision with a free electron with velocity v. The total number of collisions is obtained by integrating over all collisions with electrons with different velocities. Under most conditions the distribution of relative velocities corresponds to Maxwellian veloity dristributions at a given temperature T. Therefore in the above equation we have written $C_{l,u}(T)$.

The number of collisional de-excitations from $u \to l$ is given by:

$$n_c(u \to l) = N_u \int_0^\infty n_e(v) C'_{u,l}(v) dv = N_u n_e C_{u,l}(T) \tag{4.47}$$

Let us again deal with the simple case of thermodynamic equilibrium:

$$n_c(u \to l) = n_c(l \to u) \tag{4.48}$$

thus

$$N_l n_e C_{l,u}(T) = N_u n_e C_{u,l}(T) \tag{4.49}$$

which yields

$$\frac{N_u}{N_l} = \frac{C_{l,u}(T)}{C_{u,l}(T)} \tag{4.50}$$

and in thermodynamic equilibrium (TE) the Boltzmann formula describes the ratio N_u/N_l:

$$\frac{g_u}{g_l} e^{-(\chi_u - \chi_l)/kT} = \frac{C_{l,u}(T)}{C_{u,l}(T)} \tag{4.51}$$

Under non local thermodynamic equilibrium (NLTE) conditions we have no detailed balancing for collisional processes but also radiative excitations and de excitations. In equilibrium situations the sum of all excitation processes must equal the number of all de-excitation processes for a given energy level.

Let us consider a two level atom. The source function obtained is

$$S = (1 - \epsilon) \int_{-1}^{+1} d\mu \left(\int_{\text{line}} d\nu \phi_\nu I_\nu(\mu) \right) + \epsilon B(T) \tag{4.52}$$

ϕ_ν is the absorption profile (a Gaussian or a Voigt function) that is normalized to $\int \phi_\nu d\nu = 1$ and $\epsilon = \epsilon'/(1 + \epsilon')$ and

$$\epsilon' = \frac{C_{ul}}{A_{ul}} \left(1 - e^{-h\nu_{lu}/kT} \right) \tag{4.53}$$

C_{ul} is the rate coefficient for collisional de-excitation and A_{ul} the Einstein coefficient for spontaneous emission. If $\epsilon' \gg 1$ the collisions dominate radiative transitions and one obtains $S = B(T)$, which is local thermodynamic equilibrium LTE. Let us assume $e^{-h\nu_{lu}/kT} \ll 1$. For a resonance line $A_{ul} \sim 10^8 \, \text{s}^{-1}$. The number of collisional transitions from level u to level l is

$$n_u C_{ul} \sim n_u n_e \sigma_{ul} \bar{v}_e \tag{4.54}$$

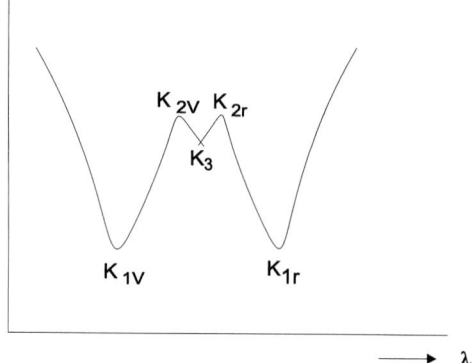

Figure 4.7: Profile of the CaII line

σ_{ul} is the cross section $\sigma_{ul} \sim \pi r_{atom}^2 \sim 10^{-15}$ cm^{-2}, the mean electron velocity $\bar{v}_e = 43\, c_s = 4 \times 10^7$ cm/s (c_s is the sound speed) and $n_e = 10^9....10^{14}$ cm^3:

$$\epsilon' = 4 \times 10^{-2}...4 \times 10^{-7} \ll 1 \qquad (4.55)$$

therefore, $\epsilon \sim \epsilon'$. The source function is dominated by the radiation field and only loosely coupled to the electron temperature via rare collisions.

By these calculation one can understand the typical profile of the Ca II H and K lines (see Fig. 4.7). There are two intensity minima on the blue and red side of the line center (called K_{1v}, K_{1r}), towards the line center two maxima (called K_{2v}, K_{2r}) and then at the line center there is a minimum (K_3). This indicates that the temperature increases in the corona. While the source function decouples from the Planck function it reaches a minimum K_1, exhibits a small maximum K_2 and finally drops towards the line center. The profile of the well known $H\alpha$ line is simpler, there is just a pure absorption. That can be explained with the structure of the H atom.

A review about the diagnostics and dynamics of the solar chromosphere can be found in Kneer and Uexküll (1999).

4.3.2 Chromospheric Heating

In the reviews of Ulmschneider et al. (1991) and Narain and Ulmschneider (1990) mechanisms which have been proposed for the heating of stellar chromospheres and coronae are discussed. These consist of heating by acoustic waves, by slow and fast MHD waves, by body and surface Alfvén waves, by current or magnetic field dissipation, by microflare heating and by heating due to bulk flows and magnetic flux emergence.

Following to Kalkofen (1990) the quiet solar chromosphere shows three distinct regions. Ordered according to the strength of the emission from the low and middle chromosphere they are

4.3. THE CHROMOSPHERE

- the magnetic elements on the boundary of supergranulation cells,
- the bright points in the cell interior, and
- the truly quiet chromosphere, also in the cell interior.

The magnetic elements on the cell boundary are associated with intense magnetic fields and are heated by waves with very long periods, ranging from six to twelve minutes; the bright points are associated with magnetic elements of low field strength and are heated by (long-period) waves with periods near the acoustic cutoff period of three minutes; and the quiet cell interior, which is free of magnetic field, may be heated by short-period acoustic waves, with periods below one minute. This paper reviews mainly the heating of the bright points and concludes that the large-amplitude, long-period waves heating the bright points dissipate enough energy to account for their chromospheric temperature structure.

Skartlien et al. (2000) studied the excitation of acoustic waves using three dimensional numerical simulations of the nonmagnetic solar atmosphere and the upper convection zone. They found that transient acoustic waves in the atmosphere are excited at the top of the convective zone (the cooling layer) and immediately above in the convective overshoot zone, by small granules that undergo a rapid collapse, in the sense that upflow reverses to downflow, on a timescale shorter than the atmospheric acoustic cutoff period (3 minutes). The location of these collapsing granules is above downflows at the boundaries of mesogranules where the upward enthalpy flux is smaller than average. An extended downdraft between larger cells is formed at the site of the collapse. The waves produced are long wavelength, gravity modified acoustic waves with periods close to the 3 minute cutoff period of the solar atmosphere. The oscillation is initially horizontally localized with a size of about 1 Mm. The wave amplitude decays in time as energy is transported horizontally and vertically away from the site of the event. They also made a prediction of how to observe these "acoustic events": a darkening of intergranular lanes, which could be explained by this purely hydrodynamical process. Furthermore, the observed "internetwork bright grains" in the Ca II H and K line cores and associated shock waves in the chromosphere may also be linked to such wave transients.

The coronal heating problem can be also studied by an energy release that is associated with chromospheric magnetic reconnection. A one-dimensional circularly symmetric supergranulation reconnection model was investigated by Roald Colin et al. (2000) with typical quiet-Sun values. In this model, the assumed source rate of elements determines heating, because all emerged elements eventually annihilate.

As an example for observational evidence we cite the paper of Ryutova and Tarbell(2000). They analyzed spectra of CII and OVI lines corresponding to chromosphere and transition region temperatures; these showed significant broadening and complex line profiles in regions overlying the sites of small scale magnetic elements in the photospheric network. Doppler shifted multiple peaks in CII line were always seen soon after the reconnection of magnetic flux tubes occurs and usually consist of supersonic and subsonic components caused by shocks propagating upward. Multiple peaks in OVI line have more diverse features: they are not as persistent as those seen in CII line, and may have the configuration of maximum intensity peaks corresponding either to forward or reflected shocks.

Ca II H_{2V} grains can also be used as indicators for shocks. Therefore spatio-temporal correlations between enhanced magnetic fields in the quiet solar internetwork photosphere and the occurrence of Ca II H_{2V} grains in the overlying chromosphere were investigated by Lites et al. (1999).

Cauzzi et al. (2000) analyzed the temporal behavior of Network Bright Points (NBPs) using a set of data acquired during coordinated observations between ground-based observatories (mainly at the NSO/Sacramento Peak) and the Michelson Doppler Interferometer onboard SOHO. The NBP's were observed in the NaD_2 line and were found to be cospatial with the locations of enhanced magnetic field. The "excess" of NaD_2 intensity in NBPs, i.e. the emission over the average value of quiet regions, is directly related to the magnetic flux density. Thus in analogy with the Ca II K line, the NaD_2 line center emission can be used as a proxy for magnetic structures.

Simultaneous CaII K-line spectroheliograms and magnetic area scans were used to search for spatial correlation between the CaII K_{2V} bright points in the interior of the network and corresponding magnetic elements and 60% of the bright points spatially coincided with magnetic elements of flux density $> 4\,\mathrm{Mxcm}^{-2}$ (Sivaraman et al. 2000).

4.3.3 Chromospheric Network, Supergranulation

On a full disk photograph taken in Ca II K a bright network surrounding darker island structures becomes visible. This pattern is known as chromospheric network. It looks like a photographic negative of the photospheric granulation pattern, however the scale is larger, typical sizes are between 20 000 and 30 000 km. This is the size of the so called supergranulation first observed by Leighton et al (1962). The bright network is cospatial with the magnetic network. The supergranulation is also visible on 30 min averaged MDI Dopplergrams. Fig. 5.4 was constructed out of a full series of 7.4 hours. The frame shown is the result of averaging 30 full disk velocity maps and subtracting the contribution from the Sun's rotation. The color scale is such that dark is motion towards the observer and bright is motion away from the observer. The signature of the waves is nearly cancelled in this image since the wave periods are mostly about 5 minutes. The resulting image clearly shows the supergranulation pattern. The "smooth" area in the center is where the supergranules do not contribute to the signal since what observers see are horizontal motions and MDI measures only the component of motion directed towards or away from SOHO (Note that supergranules are convective cells, so their motion is convective, but they are on a much larger scale than granules, and observers usually see mostly the horizontal motion).

Close inspection shows that the supergranules flow outwards from their centers so that the edges towards the center are dark (motion toward SOHO) and the edges towards the Sun's limb are bright (motion away from SOHO). These flows are about 400 m/s. The typical lifetime of a supergranular cell is about half a day. Recent investigations claim a connection between boundaries of coronal holes and supergranular structures.

In 1969 Parker and Jokipii hypothesized that the random fluid motions associated with solar supergranulation may influence the interplanetary magnetic field. Magnetic footpoints anchored in the photosphere execute a random walk and the resulting magnetic variations are carried away by the expanding solar wind. The solar satellite mission Ulysses has observed the resulting large-scale magnetic-field fluctuations in the solar wind.

By spatio-temporal averaging of two-dimensional velocity measurements obtained in the MgI 5173 line November et al. (1981) found the "mesogranulation", in order to indicate the supposed convective character of the phenomenon with a typical scale of 5 - 10 Mm and a lifetime of approximately 2 h.

The convective nature of the mesogranulation as well as the supergranulation is not sure. E.g. Rieutord et al. (2000) assign mesogranular flows with both highly energetic granules, which give birth to strong positive divergences (SPDs) among which we find exploders, and averaging effects of data processing. A similar explanation is suggested for the supergranulation.

Hathaway et al. (2000) analyzed power spectra from MDI observations. The spectra show distinct peaks representing granules and supergranules but no distinct features at wavenumbers representative of mesogranules or giant cells. The observed cellular patterns and spectra are well represented by a model that includes two distinct modes - granules and supergranules.

4.3.4 Solar Flares

The first recorded observation of a flare was a local brightening in the visible light but most solar flares can be observed in the Hα line. The typical energy release is of the order of 10^{25} J within half an hour. We only mention here, that there are stars of spectral type M where very large flares occur changing their visible luminosity but not their bolometric luminosity since energy is radiated in the IR.

Flares produce effects throughout the whole electromagnetic spectrum. They produce X rays and UV radiation which is an evidence for very high temperatures during a flare outburst. The radio waves indicate that a small fraction of the particles are accelerated to high energies. Most of the radiation is synchrotron radiation produced by electrons moving in helical paths around magnetic field lines. The flux of high energy particles and cosmic rays are also increased at the Earth as a result of an intense flare. Magnetic storms on Earth often occur with a delay of about 36 h. This is basically interpreted as an enhancement in the solar wind which compresses the magnetosphere and increases the magnetic field near the surface of the Earth. Flares occur in regions where there is a rapid change in the direction of the local magnetic field. The favored mechanism to explain the sudden energy release in flares is magnetic reconnection.

4.3.5 Classification of Solar Flares

There are different classification schemes of solar flares:

Table 4.2: Optical classification scheme of solar flares

Importance class	Area A at disk 10^{-6} sol. hemisphere
S	$A < 100$
1	$100 \leq A < 250$
2	$250 \leq A < 600$
3	$600 \leq A < 1200$
4	$A \geq 1200$

Table 4.3: Soft x-ray classification scheme of solar flares

Soft x-ray class	Peak in power of 10 in the 1-8Å flux W m^{-2}
A	-8
B	-7
C	-6
M	-5
X	-4

- Optical classification: in this scheme importance classes S,1,2,3,4 are used, according to the area of the flaring region at disk center (given in millionths of a solar hemisphere).

  In this scheme the letter S stands for subflares.

- Soft x-ray classification: since 1970 flares are also classified based on soft x-ray observations of the Sun in the 1-8 Å band by Earth orbiting satellites. The size of the flare is given by the peak intensity (on a logarithmic scale) of the emission.

  According to Table 4.3 a B5 flare has a peak flux of $5 \times 10^{-7}\,\mathrm{W m^{-2}}$. Flares smaller than C1 can only be detected during a solar minimum phase when the general x-ray background is low. Occasionally, flares exceed class X9 in intensity and are referred simply to as X10, X11...

- Classification into impulsive and gradual : in fully developed flares an impulsive phase is always followed by a gradual main phase. The classification according to the time scales is indicative of the magnetic topology.

  Long duration flares are linked to coronal mass ejections (CMEs) but recent observations also showed that some short duration flares may have ejecta. Coronal mass ejections (CMEs) leave the Sun at speeds up to 2000 km/s and can have angular spans over several active regions whereas flares imply events that are localized within a single active region. In CMEs the magnetic field lines are opened in eruptive events. There occurs a closing down or reconnection within several hours providing a prolonged energy release that

4.3. THE CHROMOSPHERE

Table 4.4: Radio classification scheme of solar flares

Type	Confined	Eruptive
Radio bursts	III/V	II/IV
Soft x-ray duration	$< 1\,\mathrm{h}$	$> 1\,\mathrm{h}$
CME	-	Yes
Interplanetary shock	.	Yes
Events/year	~ 1000	~ 10

is typical for gradual or eruptive flares. The intersection of the newly formed flare loops with the solar surface can be observed: two parallel ribbons in Hα. Therefore, in the older literature we find the designation double ribbon flares for eruptive flares.

Eruptive flares are very important because of their complexity and association with geomagnetic storms; during the event the cosmic ray intensity is also lowered.

Confined or impulsive events may also result from loop top magnetic reconnection. An impulsive flare of say 10^{24} J is typically spread over an area of several $10^{14}\,\mathrm{m^2}$ in Hα. Therefore, the main difference between eruptive and impulsive flares may be the order of intensity.

- Radio bursts and flares: solar flares are associated with radio bursts which are observed at wavelengths ranging from mm to km. The radio classification scheme was developed during the 1950s by Australian and French solar radio astronomers. The different types can be easily recognized in the so called *dynamic spectrum*: in such a diagram on the x-axis the time is plotted and on the vertical axis the frequency. Since the frequency varies with height, one can easily study the evolution with height of this phenomenon that means the propagation throughout the solar corona.

Bursts of type III and type V are characteristic phenomena of impulsive flares (or the impulsive or initial phase of fully developed eruptive flares). Type III bursts and their associated type V continua are attributed to flare-accelerated electrons moving along open field lines into the corona. Type II and type IV bursts are most commonly identified with eruptive flares. Type II bursts show a slow drift emission which can be interpreted by a shock wave moving out through the corona with a speed of $\sim 500\,\mathrm{km/s}$. Type IV emission is related to magnetic reconnection in CME.

Type II radio bursts result from plasma radiation associated with a MHD shock propagating through the corona. More than 90% of type II bursts have an associated flare. They accompany 30% of flares with an Hα importance class 2 and 3. 70% of all type II bursts are associated with a CME.

Type III bursts occur generally in active regions and thus also without flares. They are related to suprathermal electrons propagating upward in the corona. Dynamic spectra of type III bursts are characterized by the above mentioned frequency drifts from high to low frequencies as the beam excites plasma at lower densities with increasing height in the corona. The type III dm bursts are related to flares and described above.

4.3.6 Where do Flares Occur?

As a general rule, flares occur above the places in the photosphere with largest $\nabla \times B$. These are the locations where the electric current has a maximum. Preferred are regions in sunspots or groups of sunspots where new and oppositely directed magnetic flux emerges from below. Large gradual flares often occur above the neutral lines in the photosphere which separates regions with opposite magnetic polarity. Neutral lines are bridged by arcades of loops and in $H\alpha$ one sees two bright ribbons formed by the footpoints on each side of the neutral line. Flares then occur above the part of the neutral line which has experienced most shear by different surface motions on both sides. In quiet regions, the most powerful microflares occur at the boundary of supergranular cells. The frozen-in magnetic field lines are swept to the down-draft region near the supergranular boundary forming the magnetic network. At time scales of a few tens of minutes these magnetic elements can be observed to appear and disappear.

Gaizauskas (1989) made a categorization of flare precursors. According to him, a precursor is a transient event preceding the impulsive phase. We give a short list here:

- Homologous flares: these are earlier flares in the same location with similar emission patterns. They occur most often in periods of frequent flare activity. The rate of repetition ranges from a few per hour to several days.

- Sympathetic flares: these group consists of earlier flares in different locations but erupting in near synchronism. From soft x-ray images of the solar corona it is evident that there exist links between even remote active regions. Studies have shown that one flare can trigger another.

- Soft x-ray precursors: these are transient enhancements in soft x-rays lasting for several minutes; they occur in loops or unresolved kernels or close to flare sites. Weak soft x-ray bursts are often observed at the time of the onset of a CME. Sometimes several tens of minutes prior to the impulsive phase. The location is at one foot of a large coronal arch which already exists. The process can be interpreted by a small magnetic structure which interacts with the large coronal arch at one of its footpoints. The whole structure becomes then destabilized.

- Radio precursors: often tens of minutes before the onset of a flare, changes in intensity and polarity in microwaves are observed. However the correlation with flares is not very strict.

- UV precursors: small scale transient brightenings above active regions, some bright UV kernels coincide with the later flares, others do not.

- surging arches: a surging arch is a transient absorbing feature visible at wavelengths displaced from the central core of Hα. Simultaneous red- and blueshifted components are also visible. The arch is initially straight, expands and unravels in multiple strands by the time the associated flare erupts. However the link to flares is not very strong.

- Prominence eruptions: very often they precede two ribbon flares. The time delay between the onset of the prominence eruption and the impulsive phase is of the order of minutes. Enhanced mass motion, a slow rise of the prominence and untwisting can precede the main flare by hours.

Of course in all the cases joint observations covering the whole electromagnetic spectrum are important. In the review given by Aschwanden et al. (2001) the authors focussed on new observational capabilities (Yohkoh, SoHO, TRACE), observations, modeling approaches, and insights into physical processes of the solar corona.

Characteristics of flare producing sunspot groups were discussed by Ishii et al. (2000). A review about reconnection theory and MHD of solar flares is given by Priest (2000).

4.3.7 Prominences

Prominences are great areas of luminous material extending outwards from the solar atmosphere and were first observed during eclipses. They can also be observed in the light of Hα. Over the photosphere they appear as dark filaments, at the limb as bright structures. Some prominences are short lived eruptive events, others can be quiescent and survive many rotational periods of the Sun. The upper parts are often in the hot corona. Quiescent prominences are made up of material that is cooler than the photosphere. They often appear as huge arches of dense cool material embedded in the hot corona. The length of the arch is typically several 100 000 km and the height up to 10^5 km. A quiescent prominence may change into an eruptive prominence. The typical thickness of the loop is 10^4 km. At the end of its life, a prominence disperses and breaks up quietly or it becomes eruptive or matter falls back down the field lines to the photosphere. The particle densities range from $10^{16...17}$ m^{-3} which is a hundred times greater than coronal values.

A possible mechanism to understand cool prominence material (temperature about 10^4 K) is *thermal instability*. The equilibrium of the corona requires:

$$\text{heating} = \text{cooling} \tag{4.56}$$

Suppose now that this equilibrium is disturbed locally. The density of the corona increases in such a disturbed region and it will become cooler than its surroundings. If we assume that thermal conduction from the hotter surroundings cannot restore equality of temperature, the dense region will continue to cool until it reaches a new equilibrium in which its heat input balances its heat output. When a magnetic field is present, particles can only move along the field lines, this means that thermal conductivity parallel to the field lines is very much greater than $\kappa_\perp$. As a result, the longest dimension of any cool material is likely to be along the field. The equation of equilibrium of a magnetized fluid acted on by a gravitational field, g, in the z-direction is:

$$\vec{0} = -\text{grad}P - \rho g\vec{z} - \text{grad}(B^2/2\mu_0) + \vec{B}.\nabla\vec{B}/\mu_0 \quad (4.57)$$

The perfect gas law:
$$P = \Re\rho T/\mu \quad (4.58)$$

where $\Re$ is the gas constant and μ the molecular weight. In a simple model Kippenhahn and Schlüter (1957) assumed that the temperature T and the horizontal magnetic field components B_x, B_y were constant and that P, ρ and B_z were functions of x alone. The prominence is represented as a plane sheet.

4.4 The Corona

During a solar eclipse, when the moon occults the Sun for a few minutes we can observe the outer atmospheric layers of the Sun, the chromosphere and the corona the latter extending far out. The shape of the corona which extends to several solar radii depends on the sunspot cycle being more spherical around the Sun at solar maximum.

The corona includes open streamers and closed loops. These phenomena are associated with magnetic field lines. Those which return to the surface of the Sun provide closed loops, the open streamers are related to field lines which extend a large distance from the Sun carrying the solar wind, which is a continuous mass loss of the Sun. The light from the solar corona was very puzzling since many strong spectral lines could no be identified when discovered (such as Helium or Coronium; therefore their names). Later it was clarified that many of these lines are forbidden lines arising from a transition in which an electron can spend an unusually long time in an excited state before it returns to the ground level. Under normal laboratory conditions the atom will undergo many collisions and the electron will either move to the ground state without emission or move to a higher level. Therefore, no forbidden lines will be observed. In the corona and in gaseous nebulae the density of matter is extremely low, collisions are infrequent and forbidden transitions can be observed. Moreover, the coronal spectrum contains lines from highly ionized atoms indicating kinetic temperatures of several 10^6 K there which was a big surprise when discovered. Typical lines are Ca XII... Ca XV, Fe XI...Fe XV etc. Here the roman numeral is one more than the number of electrons removed from the atom. E.g. Ni XVI has lost 15 of its 28 electrons. Originally the corona could only be observed during a total solar eclipse. With a coronagraph the light from the photosphere is occulted and blocked out by a disk placed inside the telescope. Space observations allow a continuous monitoring of the corona in the UV and EUV.

The most important features seen in the corona are:

- Coronal loops are found around sunspots and in active regions in the corona. These structures are associated with the closed magnetic field lines that connect magnetic regions on the solar surface. As it is shown in the chapter on MHD, in the corona the magnetic field dominates the motion of the plasma, and therefore the plasma is aligned in magnetic loops. These loops last for days or weeks. Some loops, however, are associated with solar flares and are visible for much shorter periods. These loops contain denser material than their surroundings. The three-dimensional structure and the dynamics of these loops is investigated for that reason.

- Helmet streamers are large cap-like coronal structures with long pointed peaks. They are found usually over sunspots and active regions. Often a prominence or filament lying at the base of these structures can be seen. Helmet streamers are formed by a network of magnetic loops that connect the sunspots in active regions and help suspend the prominence material above the solar surface. The closed magnetic field lines trap the electrically charged coronal gases to form these relatively dense structures. The pointed peaks are formed by the action of the solar wind blowing away from the Sun in the spaces between the streamers.

- Polar plumes are long thin streamers that project outward from the Sun's north and south poles. At the footpoints of these features there are bright areas that are associated with small magnetic regions on the solar surface. These structures are associated with the "open" magnetic field lines at the Sun's poles. The plumes are formed by the action of the solar wind in much the same way as the peaks on the helmet streamers.

- Coronal Holes: From X-ray observations it was seen that the temperature of the corona is not uniform. The lower temperature regions are called coronal holes. They are particularly prominent near sunspot minimum and near the solar poles. Coronal holes tend to form near the centers of large unipolar magnetic regions; a comparison of the X-ray images with those of magnetic field lines calculated on the assumption that the observed photospheric field line structures extend into the corona as potential fields indicates that they are regions of open (diverging) magnetic fields. Coronal holes can also be observed in spectroheliograms taken in the 10 830 Å line of Helium. They tend to rotate more slowly than sunspots or supergranular patterns and not differentially.

The fast-speed solar wind originates form the coronal holes (e.g., Krieger et al., 1973), and accordingly they are considered the main reason for the "recurrent" type of geomagnetic activity. They may form at any latitude. For the solar cycle of greatest importance are the unipolar coronal fields. When the polar fields are strongest during sunspot minimum polar coronal holes are well defined. They disappear during the polar field reversals near sunspot maximum.

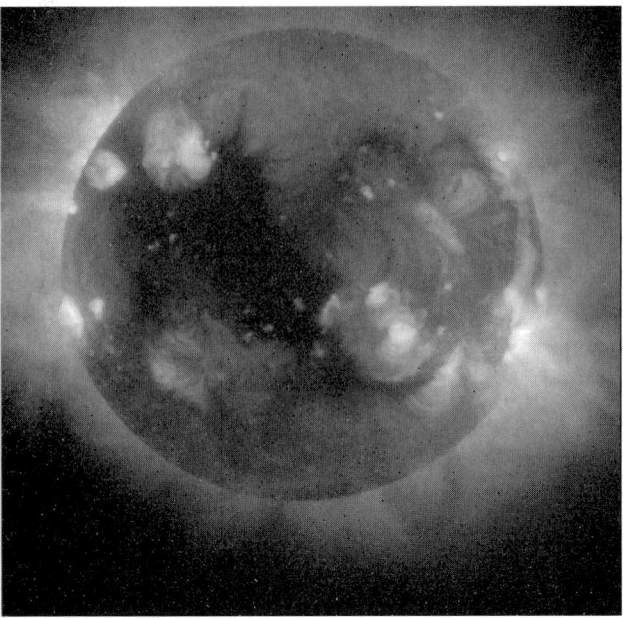

Figure 4.8: Coronal hole seen by the solar satellite YOHKOH

4.5 The Solar Wind

The Sun loses continuously mass and this mass loss is called solar wind. The existence of the solar wind was first suggested to understand magnetic storms on the Earth. During magnetic storms, the properties of the Earth's ionosphere are modified and radio communication can seriously become disrupted for some time (about 36 hours) after the observation of some violent activity on the Sun (flare). Such a perturbation cannot be caused by electromagnetic radiation from the Sun because it takes 8 minutes to reach the Earth. Therefore, it was suggested that the Sun was emitting particles which caused magnetic storms when they reach the neighborhood of the Earth.

In that context it is interesting to remark that it was Carrington who discovered in September 1859 a white light flare and then 4 hours after midnight there commenced a great magnetic storm on the magnetic instruments at Kew. He reported this observation to the Royal Astronomical Society with his comment: "while contemporary occurrence may be worth nothing, I would not have it supposed that I lean towards hastily connecting them (the event of the white light flare and the magnetic storm he observed); one swallow does not a summer make".

Another hint for solar wind arose from observations of comet tails. These are produced when comets are close enough to the Sun and the tails always point away from the Sun. Originally, it was believed that radiation pressure produces the tails. If small particles in the comet absorb radiation from the Sun they take up energy and momentum. If they subsequently emit radiation, this emission is

4.5. THE SOLAR WIND

Figure 4.9: Comet Hale Bopp (1997); the fainter ion tail is clearly seen.

isotropic into all directions and this will carry off no momentum- the matter will be pushed away from the Sun and thus the dust tails are produced. But observations showed that there is also a plasma tail consisting of ionized gas. If the Sun emits a continuous stream of plasma, the ionized solar gas would collide with atoms - momentum is transferred and charge exchange reaction occur: an electron will be exchanged between an incoming charged particles and a neutral cometary particles which produced the plasma tail. Since the charged particles move around magnetic field lines, the plasma tail is aligned with the local interplanetary field.

E.N. Parker predicted the existence of a solar wind from theoretical arguments showing that a hot corona would imply a continuous stream of plasma.

The solar wind varies in strength through the solar activity cycle. It has an average speed at the Earth of about 400 km/s. The total mass loss is a few 10^{-14} $M_\odot$/yr. This is about 1 million tons of solar material flung out into space every second. If the solar wind was the same in the past then today the total mass loss of the Sun over that period would be in the order of 10^{-4} $M_\odot$. Also, the mass loss rate is comparable with that due to nuclear reactions.

The solar wind flows along the open magnetic field lines which pass through coronal holes. Additionally to the solar wind, the Sun also looses mass by *coronal mass ejections* (CME's). Some of them but not all are accompanied by solar flares. Low speed winds come from the regions above helmet streamers we have discussed above while high speed winds come from coronal holes. However, if a slow moving stream is followed by a fast moving stream the faster moving material will interact with it. This interaction produces shock waves that can accelerate particles to very high speeds.

As the Sun rotates these various streams rotate as well (co-rotation) and produce a pattern in the solar wind much like that of a rotating lawn sprinkler. At the orbit of the Earth, one astronomical unit (AU) or about 1.5×10^8 km from the Sun, the interplanetary magnetic field makes an angle of about 45 degrees to the radial direction. Further out the field is nearly transverse (i.e. about 90 degrees) to the radial direction.

As the solar wind expands, its density decreases as the inverse of the square of its distance from the Sun. At some large enough distance from the Sun (in a region known as the *heliopause*), the solar wind can no longer "push back" the fields and particles of the local interstellar medium and the solar wind slows down from 400 km/s to perhaps 20 km/s. The location of this transition region (called the heliospheric termination shock) is unknown at the present time, but from direct spacecraft measurements must be at more than 50 AU. In 1993 observations of 3 kHz radiation from Voyagers 1 and 2 have been interpreted as coming from a radio burst at the termination shock. This burst is thought to have been triggered by an event in the solar wind observed by Voyager 2. From the time delay between this triggering event and the observation of the 3 kHz radiation, the distance of the termination shock has been put between 130 and 170 AU.

As it has been stated already, the particle density of the solar wind varies. From May 10-12, 1999, the solar wind dropped to 2% of its normal density and to half of its normal speed. This severe change in the solar wind also changed the shape of Earth's magnetic field and produced an unusual auroral display at the North Pole.

The chemical composition of the solar wind is interesting to investigate since it gives us hints about its origin, i.e. the sources. The most important fact is that the solar wind composition is different from the composition of the solar surface and shows variations that are associated with solar activity and solar features (Bochsler, 2001).

Also *magnetic clouds* have been observed in the solar wind. These are produced when solar eruptions (flares and coronal mass ejections) carry material off of the Sun along with embedded magnetic fields. These magnetic clouds can be detected in the solar wind through observations of the solar wind characteristics - wind speed, density, and magnetic field strength and direction.

References on magnetic clouds can be found in Burlaga et al. (1981). About one half of all magnetic clouds have (and usually drive) upstream interplanetary shocks, or steep pressure pulses, that in most cases possess large energy- and dynamic pressure-increases across their ramps in a stationary frame of reference. When such a sharp upstream pressure increase encounters the Earth's magnetosphere it pushes it in causing a major reconfiguration of its boundary current system measured on the ground usually some (5-10) hours before the start of the main phase of a magnetic storm (Lepping, 2001).

Coronal mass ejections (CMEs) and solar eruptions in general are assumed to result from quasi-static changes in the photospheric magnetic field which increase the magnetic energy in the corona and cause sudden release of the stored energy. This hypothesis is also called storage-release hypothesis. Chen (2001) discusses a new theory to explain the physics of CMEs. This theory claims that the initial structure is a magnetic flux rope that is ultimately connected to the solar dynamo in the convection zone and that magnetic energy propagating from the source along the submerged magnetic structure enters the corona and drives the eruption. It predicts that CMEs evolve into interplanetary magnetic clouds (MCs).

Let us give some theoretical arguments of the solar wind and describe its properties in more detail. Suppose the hot corona sits in static equilibrium on the top

4.5. THE SOLAR WIND

of the solar atmosphere. In such a case the pressure gradient in the corona must be balanced by the gravitational attraction of the Sun:

$$\frac{dP}{dr} = -\frac{GM_\odot}{r^2} \qquad (4.59)$$

In this equation we have replaced the variable M by $M_\odot$ since the mass of the corona is negligible to the total mass of the Sun. We also can write:

$$P = nkT_{\text{kin}} \qquad \rho = nm \qquad (4.60)$$

n is the number of particles per unit volume and m is the average particle mass. Please also note that T_{kin} is the kinetic temperature of the corona which is far from thermodynamic equilibrium.

In the corona, conduction is important for energy transport and if κ is the coefficient of heat conduction, then

$$\kappa = \kappa_0 T_{\text{kin}}^{5/2} \qquad (4.61)$$

where κ_0 is constant. If there is no inertial release of heat in the corona, the outward flow of heat L_{cond} must be constant:

$$L_{\text{cond}} = -4\pi r^2 \kappa_0 T_{\text{kin}}^{5/2} dT_{\text{kin}}/dr = \text{const} \qquad (4.62)$$

This equation can be integrated:

$$T_{\text{kin}}/T_c = (r_c/r)^{2/7} \qquad (4.63)$$

where r_c, T_c are radius and temperature at some point in the corona. Combining all four above equations one gets P and n as a function of r. When expanding this to the Earth one gets a kinetic temperature of 5×10^5 K and a particle density of $4 \times 10^8 \, \text{m}^{-3}$. Parker pointed out that a solution of such a system to the edge of the solar system gives nonsense. At large values of r the value of P becomes constant, so that $\rho \sim r^{2/7}$. This is higher than the pressure of the interstellar medium and thus a static model of the corona does not make sense.

If the material of the corona moves outward with a velocity v_r in the radial direction, then equation 4.59 together with 4.60 becomes

$$nmv_r \frac{dv_r}{dr} = \frac{d}{dr}(nkT_{\text{kin}}) - \frac{GnmM_\odot}{r^2} \qquad (4.64)$$

Mass conservation requires:

$$nr^2 v_r = \text{const} \qquad (4.65)$$

The thermal conduction equation must also be modified to allow for the outward flow of kinetic energy. The resulting set of equations can only be solved numerically. Let us do the following substitutions:

$$\xi = r/a$$
$$\tau = T_{\text{kin}}/T_0$$
$$\lambda = GmM_\odot/akT_0$$
$$\psi = mv_r^2/kT_0$$

Where a is the radius at the base of the corona and T_0 the value of T_{kin} there. We obtain then:

$$\frac{d\psi}{d\xi}\left[1 - \frac{\tau}{\psi}\right] = -2\xi^2 \frac{d}{d\xi}\left[\frac{\tau}{\xi^2}\right] - \frac{2\lambda}{\xi^2} \qquad (4.66)$$

4.5.1 High Speed Solar Wind

The high speed solar wind emanating from large coronal holes requires additional energy. It has been shown that Alfvén waves from the Sun can accelerate the solar wind to these high speeds. The Alfvén speed in the corona is quite large and therefore Alfvén waves can carry a significant energy flux even for a small wave energy density. These waves can therefore propagate through the corona and inner solar wind. The wave velocity amplitude in the inner corona must be 20-30 km/s. In the corona and inner solar wind region, the flow speed is much smaller than the Alfvén speed and the solar wind flow and the wave energy transport are along the magnetic field lines. In this region, the wave energy flux F in a magnetic flux tube is approximately constant:

$$F = \rho \Delta v^2 v_A A \qquad (4.67)$$

ρ... mass density, $\sqrt{\Delta v^2}$ wave velocity amplitude, v_A Alfvén speed and A is the cross section of the flow tube. The magnetic flux $\Phi = BA$ is constant, so that the wave velocity amplitude changes with density as

$$\Delta v^2 = \Delta v_0^2 \sqrt{\rho/\rho_0} \qquad (4.68)$$

The subscript 0 indicates a reference level in the inner corona.

4.5.2 Other Diagnostics for the Solar Wind

First we want to mention that besides SOHO two satellite missions measure the solar wind: Ulysses and ACE. Ulysses was launched from the space shuttle Discovery in 1990. The spacecraft made a journey to Jupiter where the giant planet's gravity pulled the spacecraft into a trajectory that carried it over the Sun's south pole in the fall of 1994 and its north pole in the summer of 1995. The next passes over the Sun's south pole occurred during fall 2000 and over the north pole during 2001. These two orbital passes provide views of the solar wind at times near the minimum of solar activity and the maximum of solar activity. The solar wind speed, magnetic field strength and direction, and composition were measured.

The Advanced Composition Explorer (ACE) satellite was launched in August of 1997 and placed into an orbit about the Lagrangian L_1 point between the Earth and the Sun. The L_1 point is one of several points in space where the gravitational attraction of the Sun and Earth are equal and opposite located about 1.5 million km from the Earth in the direction of the Sun. ACE has a number of instruments that monitor the solar wind.

The SOHO/SWAN experiment (Solar Wind ANisotropies) measures the Lα radiation that is scattered by hydrogen atoms, which flow into the solar system.

This scattered radiation is called interplanetary Lyman alpha radiation and SWAN observes interplanetary Lyman alpha radiation from all directions of the sky. These Hydrogen atoms collide with solar wind protons and get ionized. This yields to an ionization cavity around the Sun. But the form and shape of this cavity is dependent on the solar wind. Therefore the measurement of the interplanetary UV Lα glow permits to determine the solar wind latitudinal distribution. If the solar wind were isotropic, the hydrogen distribution and the Lyman alpha emission pattern would be axisymmetric around the direction where the interplanetary hydrogen flows into the solar system. However, this is not true.

Planetary Magnetospheres

Here we briefly describe measurements of the magnetic fields of other planets which are useful as diagnostics of the solar wind. The magnetic field of Mercury and the structure and dynamics of Mercury's magnetosphere are strongly influenced by the interaction of the solar wind with Mercury. In order to understand the internal magnetic field, it will be necessary to correct the observations of the external field for the distortions produced by the solar wind. The satellites Helios 1 and 2 made a number of passes in the region traversed by the orbit of Mercury; thus it is possible to investigate the solar wind environment of Mercury. The variables that govern the structure and dynamics of the magnetospheres of Mercury and Earth are approximately 5-10 times larger at Mercury than at Earth. Thus, the solar wind interaction with Mercury will be much stronger than the interaction with Earth (Burlaga, 2001). The solar wind is not constant and since Mercury is closer to the origin of it, the solar wind at Mercury is probably more variable than that at Earth.

Mercury, Earth, Jupiter, Saturn, Uranus, Neptune, and Ganymede (satellite of Jupiter), have presently-active internal dynamos while Venus, Mars, at least two of the Galilean moons, the Earth's moon, comets and asteroids do not. These active dynamos produce magnetic fields that have sufficient strength to stand off the pressure of the exterior plasma environment and on the other hand interesting interactions with the solar wind can be studied. Moreover, e.g. the jovian magnetosphere includes a strong time-varying energy source that adds to the dynamics of its magnetosphere and produces a quite different circulation pattern than that found at Earth and, presumably, Mercury. Also the unmagnetized planets Venus, Mars and even comets have induced magnetospheres associated with the solar wind interaction with their atmospheres. Cometary magnetospheres, parts of which can be remotely sensed, exhibit spectacular disruptions called tail disconnections. Even the atmosphereless bodies with weak magnetic fields can interact with the solar wind. Small magnetic anomalies on the moon and possibly asteroids cause weak deflections of the solar wind. This is discussed in the paper of Russell (2001).

Krymskii et al. (2000) investigate the interaction of the interplanetary magnetic field and the solar wind with Mars. Data from the Mars Global Surveyor mission have shown that localized crustal paleomagnetic anomalies are a common feature of the Southern Hemisphere of Mars. The magnetometer measured small-scale magnetic fields associated with many individual magnetic anomalies (magnitudes ranging from hundreds to thousands nT at altitude above 120 km). Thus Mars is globally different from both Venus and Earth. The data collected by Lunar Prospector near the Moon were interpreted as evidence that above regions of inferred strong surface magnetic fields on the Moon the solar wind flow is deflected, and a small-scale mini-magnetosphere exists under some circumstances. With a factor of 100 stronger magnetic fields at Mars and a lower solar wind dynamic pressure (because of the greater distance), those conditions offer the opportunity for a larger size of small 'magnetospheres' which can be formed by the crustal magnetic fields. The Martian ionosphere is controlled both by solar wind interaction and by the crustal magnetic field. Therefore, the nature of the Martian ionosphere is probably different from any other planetary ionospheres, and is likely to be most complicated among the planetary ionospheres (Shinagawa, 2000).

Bochsler (2001) discusses the effect that CMEs have on the composition of the particle flux. SOHO/LASCO observations show that even at times of solar minimum these spectacular events fed an important part of the low latitude corona. Elemental and isotopic abundances determined with the new generation of particle instruments with high sensitivity and strongly enhanced time resolution indicate that mass-dependent fractionation can also influence the replenishment of the thermal ion population of the corona. Furthermore, selective enrichment of the thermal coronal plasma with rare species such as ^{3}He can occur. Such compositional features in energetic particles are known from impulsive flare events but it seems that also the above mentioned effect must be taken into account.

The global solar wind structure from solar minimum to solar maximum is reviewed by Gibson (2001).

4.6 Heating of the Corona

The original idea for the heating of the corona was entirely non-magnetic. From laboratory experiments we know that if a fluid is set into violent motion, it emits sound with the amount of sound rising as a high power of the average velocity of the fluids. As we have seen, the outer layer of the Sun has convective motions. If these convective motions produce sound waves, they must propagate outwards from the surface of the Sun. The wave motion has an energy density of

$$E_{\text{wave}} = \frac{1}{2}\rho v^2 \qquad (4.69)$$

This energy is conserved. If the wave moves into a region of lower density, then the wave amplitude must increase. The wave turns into a shock wave and there is a strong dissipation of energy. This is converted into heat and the local temperature raises. However it turned out that a purely acoustic heating of the corona is not sufficient to explain the high temperatures there. Acoustic heating may be important in the outer layers of some stars.

Today we assume that the following two processes are the main reason for the hot corona:

- MHD waves: as it has been outlined, when a magnetic field is present there are two characteristic speeds of wave propagation, the sound speed c and the Alfvén speed c_H. If $c_s \gg c_H$ magnetic effects are negligible but this is not the case for the outer solar atmosphere. The heating process by MHD waves is analogous to the above mentioned acoustic heating. But it has to be noted again that MHD waves have an anisotropic propagation.

- Magnetic reconnection: The footpoints of magnetic fields often are seen to be anchored in the photosphere. In this region they are being continually moved around by convective motions. Thus magnetic reconnection occurs and electric currents flow which are dissipated.

4.7 Variations of the Solar Diameter

When measuring the solar diameter one has to take into account that the Sun is a gaseous sphere and its diameter is in principle a matter of definition. When looking at the solar limb, the decrease of the tangential optical depth from unity to essentially zero occurs over only a few hundred km which is small compared to the total solar radius. A major decrease occurs within 0.2 arcsec of both sides of the point of inflection when regarding a scan. Therefore, one can define a solar diameter this way.

Ground based measurements of the solar diameter exist over more than 300 years. Because of the small variations the results are controversial and inconsistent. The first determination of the solar diameter was made by Aristarchus 270 b.c. He obtained a value of 900 arcsec. The first accurate measurements were performed by Mouton in the year 1970 at Lyon during the period of 1959-1961 and he obtained a value of 960.6 arcsec for the solar semidiameter. From historical data it may be deduced that the solar radius may have been larger during the Maunder Minimum. As we have seen this minimum of solar activity coincided with extremely cold periods in Europe and the Atlantic regions (Ribes et al. 1991). Also Laclare et al. (1996) found a larger solar radius during solar minimum. However other groups (Ulrich and Bertello, 1995, Noel, 1997 and Basu 1998) found a positive correlation: the solar diameter increases with enhanced solar activity. Besides a possible variation of the solar radius with the solar cycle there are also hints that the solar radius changes over timescales of 1 000 days to 80 years (Gilliland, 1980). Thus we see that there is a wide range of measurements and the results are ambiguous. From helioseismic measurements Dziembowski et al. (2000) deduced solar radius fluctuations and they found a change of 10 mas between 1996 and 1998.

The fact that these measurements are controversial is related to the problem that the fluctuations are quite small and Earth bound observations are always limited by seeing. Thus one wants to reduce this effect by using balloon borne instruments (Sofia et al. 1994) or satellite data (Michelson Doppler Imager, MDI on board of SOHO). These data are free of atmospheric disturbances and promise very accurate determinations of the solar radius.

Why is it important to study solar radius variations? The radiated energy of the sun comes from the nuclear energy generation (Fusion of H to He) in the deep solar interior. In the solar core at a temperature of more than 10 Million K the energy is generated by the fusion of H to He and high energetic γ ray photons are emitted. These energetic photons are absorbed and re-emitted in the solar interior (mean free path between the absorption processes is only 1 cm) and therefore for a photon generated by such nuclear reaction, it takes more than 1 million years to diffuse out of the core region. Thus one can argue that the emergent luminosity at the core outer boundary is effectively constant on solar cycle timescales. If there is any luminosity variability at the surface there must be an intermediate energy reservoir between the core and the photosphere. There are several mechanisms for storing energy during a solar activity cycle, such as magnetic fields or gravitational energy. Each of them leads to distinct perturbations in the equilibrium structure of the sun. Therefore, one can argue that a sensitive determination of the solar radius fluctuations can help to understand the solar cycle and it is clear that the magnitude of the radius fluctuations compared to the luminosity change contains information on where and how energy is stored.

Sofia and Endal (1979) introduced the parameter W by:

$$W = \frac{\delta r}{r} / \frac{\delta L}{L} \qquad (4.70)$$

The models predict a wide range for W:

- 2×10^{-4} Spruit, 1982;
- 8×10^{-4} Gilliland, 1980;
- 5×10^{-3} Dearborn and Blake, 1980;
- 7.5×10^{-2} Sofia and Endal, 1979;
- W could be positive or negative, Lydon and Sofia (1995).

In the following we discuss briefly some measurement methods and give the results.

4.7.1 Satellite Measurements

Emilio et al. (2000) used SOHO MDI measurements to derive possible variations of the solar diameter. They used 1 minute cadence images, and these were low pass filtered in order to remove solar 5 minute p mode intensity oscillations. The limb pixels (2 arcsec / pixel) were downlinked every 12 minutes. The data set used was between 1996 April 19 and 1998, June 24. They did not use data obtained after the recovery of SOHO in November 1998 because of the frequent instrument mode interruptions and focal length calibration difficulties. They find annual radius variations at an amplitude of 0.1 arcsec and a secular increase of about the same amplitude over the period between 1996 and 1998. The systematic variation is caused by the changing thermal environment of the MDI front window which yields small but measurable changes in the telescope focal length. A temperature gradient of a few degrees from the center of the window to the aluminium cell at the filter edge can produce a weak lens effect; that corresponds to a focal length of a few km and changes the telescope focal length by a few parts in 10^4. The secular change is also influenced by the degradation of the front window and increased absorptivity.

Thus the MDI data yield lower values of opposite sign. Since Sofia et al. (1979) claimed that $W \sim 0.075$, solar cycle changes which affect the convective efficiency near the photosphere will have a large effect on the solar radius; the MDI measurements rule out this high value of W and suggest that solar cycle luminosity changes are not caused by superficial fluctuations in the outer layers of the Sun.

4.7.2 Measurements with an Astrolabe

Laclare et al. (1996) published results of solar diameter measurements obtained with the Danjon astrolabe at the Observatoire de la Côte d'Azur; this program was initiated in 1975 and the instrument consists of a set of 11 reflector prisms which enables the measurement of the diameter up to 22 times a day at different zenith distances (from 30 to 70 degrees). Observing a transit requires the recording of the time when both images of the Sun's edge, i.e. the direct one and its reflection on a mercury surface become tangent to each other. At this instant the Sun's edge crosses the parallel of altitude (almucantar) which is defined in the instrument by the angle of the reflector prism and also by the refraction and other terms.

Of course this technique requires a true stability of the almucantar during observation and Zerodur types of ceramic reflector prims (which are practically unaffected by dilation) and a mercury mirror establishing the horizontal plane are used.

Furthermore, the observations were cleared of personal bias by using an acquisition system equipped with a CCD camera at the focal plane of the instrument. The limb was defined as the point where the intensity distribution on a CCD line has its inflection point (zero of the second derivative of the solar limb function). For each frame then the limb was reconstructed by a least-square adjustment of a parabola through the inflection points.

The mean value of the semi diameter was obtained by visual measurements and by the above described data acquisition system:

- 5 000 visual measurements, same observer; 1975-1994: 959.46±0.01 arcsec., broad band (200 nm) filter was used centered on 540 nm.

- CCD acquisition program: 981 CCD measurements in the period 1989-1994; mean value= 959.40±0.01 arcsec.

- Correlations with solar activity: nearly opposing trend; high activity means smaller diameter.

It is important to notice that all ground based observations must take into account the quasi biennial oscillation in the Earth's atmosphere.

Table 4.5: Solar Diameter Measurements

Author	Period	Value	Corr. coeff.
Wittmann	1972-1991	+0.25 arcsec	0.9
Laclare	1978-1994	+0.09	0.4
Leister	1980-1993	+0.09	0.2
Kubo	1970-1991	+0.05	0.8
Bode	1976 1994	0.00	0.1
Neckel	1981-1990	0.00	0.1

4.7.3 Other Semi Diameter Variations

In this paragraph we just summarize some semi diameter variations given in the literature. The authors do not suggest a possible solar cycle dependence.

Other astrolabe measurements were done by Sanchez et al. (1995) at the San Fernando Observatory (Cadiz), Noel (1995) at Santiago and Leister et al at Sao Paulo Observatory (1990). Ribes et al (1991) report on photoelectric measurements made at HAO in Boulder from 1986-1990; Wittmann et al. (1993) report on measurements using drift timing in Izana and Locarno. Other authors measured the solar diameter from eclipse data (e.g. Kubo, 1993).

Chapter 5

Testing the Solar Interior

So far we have discussed several aspects of solar activity. In order to understand the solar activity cycle it is necessary to test the solar interior and there are two observational possibilities:

- Solar neutrinos, that are emitted during nuclear reactions.
- Propagation of seismic waves.

5.1 Neutrinos

5.1.1 General Properties

Let us consider the well known Beta decay. If the nucleus of an atom has too many neutrons the most likely course is that the nucleus emits an electron. This has the same effect as turning one of the neutrons into a proton. Such electrons are historically referred to as beta rays having been named before they were identified as electrons. An example of beta decay is the decay of tritium or ^{3}H into ^{3}He:

$$^3\text{H} \rightarrow\, ^3\text{He} + e^- \qquad (5.1)$$

Interestingly enough, the neutrino was first invented as an ad hoc hypothesis, in order to save the laws of conservation of energy and momentum from falsification. Around 1930, in the first detailed studies of radioactive beta-decays, it was found that some energy and momentum went missing in each decay. Beta decay involves the conversion of a neutron into a proton, accompanied by the emission of an electron, and nothing else visible. The energy carried away by the electron ought to match the energy released by the atom in the process but it didn't! Wolfgang Pauli proposed to explain this discrepancy by postulating that an additional, invisible particle was emitted along with the electron, carrying away the missing energy and momentum. This "ghost particle" was named *neutrino*.

There exist three families of elementary particles, each family consisting of two quarks, and two leptons. Quarks are constituents of protons and neutrons. Lepton is the collective term for electrons and neutrinos and their relatives in the other families. The electron and the (electron-)neutrino ν_e make up the lepton pair of the first family. In the other two families, the electron-equivalents are called muon μ and tau τ, each with their neutrino partner, called mu-neutrino ν_μ and tau-neutrino ν_τ. So we have three different charged leptons: electron, muon, and tau; and three neutrinos, one associated with each of the three charged leptons. The ν_τ was discovered in 1988.

5.1.2 Solar Neutrinos

As we have seen, neutrinos are produced in the first reaction of the pp chain having an energy between zero and 0.42 MeV. The maximum energy for the neutrinos from the decay of ^{8}B is about 15 MeV. All neutrinos interact very weakly with matter, the probability of absorption increases with their energy. The rare ^{8}B neutrinos are more likely to be absorbed. The absorption cross section is the effective area offered by a target particle to a beam of incident particles. For neutrinos the cross section σ is:

$$\sigma \sim 10^{-50}\,\text{m}^2 \tag{5.2}$$

When we compare this value to the cross sections in atomic and nuclear physics which are about $10^{-20}...10^{-30}\,\text{m}^2$ we see that neutrinos can penetrate the whole Sun without being absorbed and therefore they can be used to test our models. The distance between collisions, the mean free path l, if the target particles have a number density $n\,\text{m}^{-3}$, is given by:

$$l = \frac{1}{n\sigma} \tag{5.3}$$

For a solid target one has $n \sim 10^{29}$ and therefore $l \sim 10^{21}\,\text{m}$ for neutrinos. So neutrinos have an extremely large value of the mean free path. How can we detect them? There are many neutrinos coming from the Sun passing the Earth: about $10^{15}\,\text{m}^{-2}\text{s}^{-1}$. With the cross section and the number density given above, the number of detections N would be:

$$N \sim 10^{-6}\,\text{m}^{-3}\,\text{s}^{-1} \tag{5.4}$$

That means about one neutrino per month per cubic meter of the detector.

5.1.3 Solar Neutrino Detectors

The first experiment to detect solar neutrinos was a ^{37}Cl reaction with neutrinos resulting in ^{37}Ar which is unstable and decays to ^{37}Cl. The decay of Ar can be detected.

$$\begin{aligned} ^{37}\text{Cl} + \nu_e &\rightarrow\ ^{37}\text{Ar} + e^- \\ ^{37}\text{Ar} &\rightarrow\ ^{37}\text{Cl} + e^+ + \nu_e \end{aligned} \tag{5.5}$$

5.1. NEUTRINOS

Only neutrinos with energies > 0.8 MeV can be detected by this reaction. This rules out the most numerous low energy neutrinos (first reaction in the pp chain). The ^{37}Cl is in a tank containing 10^5 gallons of C_2Cl_4 perchlorethylene in the Homestake Gold Mine in Lead, South Dakota. The experiment has to be placed deep below the surface to avoid contaminating reactions produced by cosmic rays. Ar is an inert gas, one can extract it from the tank and observe its decay elsewhere.

Neutrino detections are measured by the solar neutrino flux unit defined by:

$$1\,\text{SNU} = 10^{-36}\,\text{interactions}\,\text{s}^{-1}\,\text{target atom}^{-1} \tag{5.6}$$

Since the experiment contains about 2^{30} ^{37}Cl atoms one has to expect one detection every 5×10^5 s.

Theoretical models of the Sun predict the following count rates:

$$\text{Bahcall, Pinsonneault (1992)}\ 8.0 \pm 3.0\,\text{SNU}$$
$$\text{Turck-Chièze, } Lopez\ (1993)\ 6.4 \pm 1.4\,SNU \tag{5.7}$$

However the measured flux is:

$$2.28 \pm 0.23\,\text{SNU} \tag{5.8}$$

As it has been stressed already, the chlorine experiment is (according to standard solar model predictions) sensitive primarily to neutrinos from the rare fusion reaction that involves ^{8}B neutrinos which are produced in only 2 of 10^4 terminations of the pp cycle. In a conference held in Brookhaven, 1978, it was therefore suggested to design new experiments that are sensitive to the low energy neutrinos from the fundamental pp reaction. Using ^{71}Ga instead of chlorine was first proposed by the Russian theorist Kuzmin in 1965. However about 3 times the world's annual production of Ga would be needed to perform that experiment.

In the Ga experiment, neutrinos with an energy ≥ 0.2332 MeV can initiate the reaction:

$$\nu_e + ^{71}\text{Ga} \rightarrow e^- + ^{71}\text{Ge} \tag{5.9}$$

Therefore, many of the pp neutrinos are included. The SAGE is a Russian/American experiment and uses 60 tons of metallic Gallium. The GALLEX experiment is a European experiment located underground in Italy. It uses 30 tons of Ga in a $GaCl_3HCl$ solution. More than half of the neutrinos that can be detected with this experiment come from the second most important contribution of the pp chain, from ^{7}Be. Again the results are inconsistent with theoretical predictions with a discrepancy by a factor of about 2 however they provided a first experimental indication of the presence of pp neutrinos.

The Kamiokande experiment uses a large tank of pure water sited underground and its aim was to study the possible decay of the proton. The half life of a p is $\sim 10^{30}$ yr. The neutrino detector picked up a number of neutrinos from the explosion of the supernova SN 1987A in the Large Magellanic Cloud, which is a neighbor of our galaxy. In an updated version (Kamiokande II) 0.68 kilotons of water were used and ^{8}B above 7.5 MeV can be detected. The water experiment Kamiokande detects higher energy neutrinos (above 7 MeV) by neutrino-electron scattering ($\nu+e \to \nu'+e'$) and according to the standard solar model the ^{8}B decay is the only important source of these higher-energy neutrinos. The experiment clearly showed that the observed neutrinos come from the sun because the electrons that are scattered by the incoming neutrinos recoil predominantly on the direction of the sun-earth vector. The relativistic electrons are observed by the Cerenkov radiation they produce in the water detector.

The results of the gallium experiments, GALLEX and SAGE gave an average observed rate of 70.5 ± 7 SNU. This is in agreement with the standard model by the theoretical rate of 73 SNU that is calculated from the basic pp and pep neutrinos. The ^{8}B neutrinos which are observed above 7.5 MeV in the Kamiokande experiment, must also contribute to the gallium event rate. This contributes another 7 SNU, unless something happens to the lower energy neutrinos after they are created in the Sun. Thus the Ga experiments are in accordance with predictions if we exclude everything but the pp neutrinos. This is sometimes called the third neutrino problem.

The calculated pp neutrino flux is approximately independent of solar models; it is closely related to the total luminosity of the sun.

Summarizing the the neutrino problem we can state:

- smaller than predicted absolute event rates in the chlorine and Kamiokande experiments.

- incompatibility of the chlorine and Kamiokande experiments,

- very low rate in the Ga experiment which implies the absence of ^{7}Be neutrinos although ^{8}B neutrinos are present.

Solar neutrino experiments are currently being carried out in Japan (Super Kamiokande,Takita (1993), Totsuka (1996)), Canada (SNO, Sudbury, using 1 kiloton of heavy water; Mc Donald, (1991)) and in Italy (BOREXINO, ICARUS, GNO (Gallium Neutrino Observatorium), each sensitive to a different energy all working in Gran Sasso, Arpesella et al. (1992)), in Russia (SAGE, Caucasus) and in the United States (Homestake). The SAGE, chlorine and GNO work radiochemical, the others electronic (recoil electrons produced by the neutrino interactions using Cerenkov effect).

The gallium solar neutrino detector consists of 30.3 tons of gallium, in the form of 103 tons of an acidic solution of gallium chloride.

Solar neutrinos with energy larger than 230 keV interact with the ^{71}Ga nuclei. As a consequence the following reaction takes place (inverse beta decay):

$$^{71}\text{Ge}(\nu, e^-)^{71}\text{Ga} \tag{5.10}$$

It is important to remark that, due to the low threshold for this reaction, most of the captured solar neutrinos are the low energy 'pp' neutrinos.

The neutrino interaction rate is very low, of the order of one interaction per day in the whole detector mass.

^{71}Ge produced by neutrinos is radioactive (halflife about 16 days), and decays by electron capture into ^{71}Ga (the reverse process of the solar neutrino capture).

5.1. NEUTRINOS

The ^{71}Ge accumulates in the solution, reaching equilibrium when the number of ^{71}Ge atoms produced by neutrino interactions is just the same as the number of the decaying ones. When this equilibrium condition is reached, about a dozen ^{71}Ge atoms are present inside the 103 tons gallium chloride solution (containing 10^{29} Ga nuclei). The solution is exposed for four weeks, about 12 Ge nuclei are present and are chemically extracted into water by pumping 3000 m^3 of N through the tank. Then it is converted into Ge gas and the decays are observed.

The Sudbury Neutrino Observatory (SNO) is located at 6800 feet under ground in a mine in Sudbury, Ontario. The neutrinos react with heavy water producing flashes of light (Cerenkov radiation). The principle is as follows:

$$\nu_e + d \to p + p + e^- \tag{5.11}$$

As the neutrino approaches the deuterium nucleus d a heavy charged particle of the weak force (called the W boson) is exchanged. This changes the neutron in deuterium to a proton, and the neutrino to an electron. The electron, according to mechanics, will get most of the neutrino energy since it has the smaller mass (consider a gun that is fired; the bullet, being lighter, gets most of the energy). Due to the large energy of the incident neutrinos, the electron will be so energetic that it will be ejected at light speed, which is actually faster than the speed of light in water. This causes the optical equivalent of a "sonic boom", where a "shock wave of light" is emitted as the electron slows down. This light flash, called Cherenkov radiation, is detected.

The current status of solar neutrino experiments was reviewed by Suzuki (1998).

5.1.4 Testing the Standard Solar Model

We will speak about helioseismology in a later chapter, however in this context it should be noted that results from helioseismology have increased the disagreement between observations and the predictions of solar models with standard neutrinos. Helioseismological measurements demonstrate that the sound speeds predicted by standard solar models agree with high precision with the sound speeds of the sun inferred from measurements. This leads to the conclusion that standard solar models cannot be wrong to explain the discrepancy.

The square of the sound speed is:

$$c^2 \approx T/\mu \tag{5.12}$$

where T is the temperature and μ the mean molecular weight. Sound speeds can be determined with the aid of helioseismology to a very high accuracy (better than 0.2% rms throughout nearly the whole sun). Thus one can estimate tiny errors in the model values of T and μ as measurable discrepancies in the precisely determined helioseismological sound speed:

$$\frac{\delta c}{c} \simeq \frac{1}{2}\left(\frac{\delta T}{T} - \frac{\delta \mu}{\mu}\right) \tag{5.13}$$

The quantitative agreement between standard model predictions and helioseismological observations rules out solar models with temperature or mean molecular weight profiles that differ significantly from the standard values. This observational agreement rules out in particular solar models in which deep mixing has occurred. The best agreement is obtained when including the effect of particle diffusion-selective sinking of heavier species in the sun's gravitational field. Models without taking into account of this effect have rms discrepancies between predicted and measured sound speeds as large as 1% (e.g. Turck-Chièze and Lopez (1993) whereas models including this effect have rms discrepancies of 0.1% (Bahcall et al., 1997).

The sound-speed profile in the Sun was determined by carrying out an asymptotic inversion of the helioseismic data from the Low-Degree (l) Oscillation Experiment (LOWL), the Global Oscillation Network Group (GONG), VIRGO on SOHO, the High-l Helioseismometer (HLH), and observations made at the South Pole (Takata and Shibahashi, 1998). Then the density, pressure, temperature, and elemental composition profiles in the solar radiative interior were deduced by solving the basic equations governing the stellar structure, with the imposition of the determined sound-speed profile and with a constraint on the depth of the convection zone obtained from helioseismic analysis and the ratio of the metal abundance to the hydrogen abundance at the photosphere. Using the resulting seismic model, neutrino fluxes were estimated and the neutrino capture rates for the chlorine, gallium, and water Cerenkov experiments. The estimated capture rates are still significantly larger than the observation.

Solar models with helioseismic constraints and the solar neutrino problem are discussed in Watanabe and Shibahashi (2001) and Roxburgh (1998).

Is there a correlation between neutrino fluxes and solar activity? On the basis of an analysis of the ^{37}Ar production rate at the Homestake station for the period 1970-1990, Basu (1992) found that the solar neutrino flux varies with time in proportion to the solar wind flux. However, Walther (1999) found that there exists no significant correlation between the Homestake neutrino data up to run 133 and the monthly sunspot number, according to a test that is based on certain optimality properties for this type of problem. It is argued that priorly reported highly significant results for segments of the data are due to a statistical fallacy.

5.1.5 Solution of the Neutrino Problem

How to explain this discrepancy between observations and theory? One explanation comes from particle theory itself. There are three conserved quantities, called electron, muon and tauon lepton numbers and correspondingly three types of neutrinos. There are however indications that some modifications to this standard model are required. These involve that the neutrinos have small masses and that the neutrinos can transform from one type to another. The Mikheyev-Smirnov-Wolfenstein effect (MSW) explains such neutrino oscillations and by the above mentioned experiments we can only detect electron neutrinos. Another explanation of the discrepancy is that the flux is variable during the solar cycle. This might be explained if the neutrinos possess a magnetic moment and if interaction with the solar magnetic field is possible.

Let us give a very simplified explanation of neutrino oscillations. An indispens-

5.1. NEUTRINOS

able, but counterintuitive, concept in quantum mechanics is that of superposition. Suppose a certain particle has a property that can have several different values; the classic example is that of Schrödinger's cat (look it up!), let us consider a more practical one: ordinary playing cards have the property 'suit', with the four possible values 'spades', 'hearts', 'diamonds', and 'clubs'. In ordinary non-quantum life, each individual card has a well-defined *suit*. However a quantum card may be in a mixed state, a superposition of, say 30% spades, 60% hearts, and 10% clubs. When you check which suit that card belongs to, you have a 30% chance of finding that it's a spade, 60% chance of finding it's a heart, and so on. Note that this is not just a matter of your ignorance of the card's "true" suit the point is, it doesn't have a single well-defined suit until you check it.

In particle physics, the equivalent of the suits are the three families, discussed above. A neutrino may belong to any one of the three families, making it ν_e, or a ν_μ, or a ν_τ. Or, it may be a superposition of the three family flavors, mixed in some proportions. Now, the standard model assumes that the neutrinos emitted from the sun are in a pure ν_e state, without mixing. This can be understood with the quantum mechanical concept of eigenstates. This is well known for the K meson. An eigenstate is a state that is recognized as pure, non-mixed, without superposition, in a certain context. In quantum mechanics different interactions recognize and interact each with a different set of eigenstates for the particles. Try to apply this to a card game. In different games a heart would become a spade etc. For most particles and interactions the different eigenstates are identical. This is not the case for the weak interaction. The weak eigenstates of quarks are different from their strong/electromagnetic eigenstates. The K0 mesons are produced in strong interactions of quarks, but decay through weak interactions of their constituent quarks. Thus, the production eigenstates are different from the travel/decay eigenstates of the K0. As far as the weak interaction is concerned, leptons are expected to behave in the same manner as quarks. If neutrinos do have a tiny mass, and different neutrinos have different masses, they will behave in the same way as K0 mesons. They will be produced in a weak-interaction eigenstate, but travel in a mass eigenstate. The mass eigenstate may be different from the weak eigenstate. The weak-interaction eigenstates are the three neutrino flavors discussed earlier: ν_e, ν_μ, ν_τ. When they arrive and interact with our detectors, they do not arrive as the original weak eigenstate in which they were produced, but as a mixture of two or more flavors. This is a potential solution to the solar neutrino problem, since the experiments measure an apparent disappearance of electron-neutrinos, without measuring the other flavors. If the neutrinos oscillate from the 100% ν_e that they are produced as in the sun, to a mixture with around 40% ν_e electron-neutrino and 60% some other neutrinos, we get an agreement with experimental data.

Neutrino oscillations and the solar neutrino problem are discussed by Haxton (2001).

The search for neutrino decays during the 1999 solar eclipse is discussed in Cecchini et al. (2000) involving the emitted visible photons, while neutrinos travel from the Moon to the Earth.

Alternate Solar models

Other suggestions to solve the neutrino problem are:

1. There is an additional force resisting gravity in the solar interior which reduces the central pressure and temperature- maybe rapid rotation, strong internal magnetic fields. Since the pp chain is strongly dependent on temperature, this might explain a different SNU.

2. The Sun contains a central black hole or neutron star. There occurs a gravitational release from accretion providing much of the radiated energy.

3. The surface chemical composition is not typical of the interior composition.

4. Waves or weak interacting particles contribute to the energy transport.

Today the most likely solution of the neutrino problem are the neutrino oscillations suggesting that our solar model is quite correct. Bahcall and Davis (2000) gave a recent review about the solar neutrino problem and suggest further experiments.

5.2 Helioseimology-Solar Oscillations

In the 1960s the five minutes oscillations were detected on the solar surface. These are vertical oscillations with a strongly varying amplitude but a period of five minutes, the maximum velocities about 0.5 km/s towards or away from the observer. The pattern persisted for about half an hour (six cycles of wave motion), then faded away but a similar pattern would then be in progress elsewhere.

It was realized that these oscillations could be understood in terms of a superposition of many normal modes of solar oscillations.

Let us consider one analogy: For seismic waves on the Earth one usually has only one source of agitation - an earthquake. For the Sun, there are many sources of agitation of solar "seismic" waves; these sources of agitation causing the solar waves are processes in the larger convective zone. Because there is no single source, we can treat the sources as a continuum, so the ringing Sun is like a bell struck continuously with many tiny sand grains.

Waves travelling from the interior of the Sun up to the surface would be reflected back again at the surface boundary. Imagine a wave normal to the surface of the Sun and travelling towards the center. As the wave travels deeper into the interior, the temperature increases and the wave is progressively refracted away from the normal until it turns around and returns to the surface. At the Sun's surface the sharp density gradient causes subsequent reflection and the wave heads back into the Sun. Thus the Sun is a *resonant cavity*, standing waves are created. The more often a wave returns to the surface, the less deeply it penetrates before being turned back and conversely, waves reflected only a few times from the surface probe much deeper into the Sun .

There are two different types of oscillations depending on the restoring force.

- p- modes: the restoring force is the pressure;

5.2. HELIOSEIMOLOGY-SOLAR OSCILLATIONS

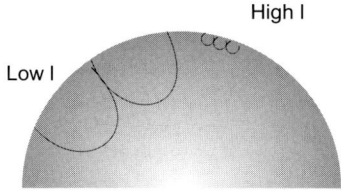
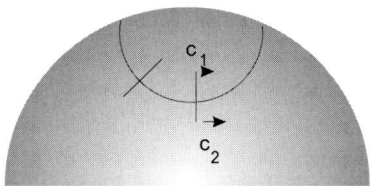

Low l modes are reflected deeper
high l modes are reflected higher

Velocity of sound c depends on temperature T. T increases inwards, therefore the wavefront will be reflected.

Figure 5.1: Left: waves with low and high l; the low l modes are reflected deeper than the high l modes. Right: Explanation how the waves are reflected in the solar interior. The wavefront (normal to the propagation) is deflected since the sound velocity is higher in deeper layers ($c_2 > c_1$).

- g-modes: the restoring force is the gravity

There exist also surface waves which are called f-modes. The p-modes have frequencies between 1 hour and two minutes and include the five minutes oscillations discussed above. The g-modes have much longer periods than the p-modes. It can be shown that they are trapped in the solar interior beneath the convection zone. The energy generated in the sun is first transported by radiation and then at a depth of about 200 000 km by convection. In this convection zone the amplitudes of the g-modes are damped exponentially and thus it is extremely difficult to observe them at the solar surface.

How can we describe the solar oscillations? First we must make some simplifications. We assume that the sun is strictly spherical. This will provide a spectrum of oscillation frequencies which will be modified by a) rotation and b) magnetic fields. A second approximation is that the oscillations are adiabatic. This approximation is valid since the oscillation period is in general much smaller than the relevant thermal timescale. A third approximation is that we neglect a change of the gravitational field of the Sun during the oscillation. This is not true for radial oscillations: in radial oscillations all matter at any solar radius moves inward or outward in phase. However if we consider nonspherical modes at short wavelengths in the horizontal direction this is again a good approximation.

Any oscillation can be described by introducing three quantum numbers n, l, m. The meaning of these numbers is as follows:

- n denotes the number of points in the radial direction at which the amplitude of the oscillation vanishes.

- l, m determine the angular behavior of the oscillation over the surface of the Sun. In addition we have the relation $-l \leq m \leq +l$.

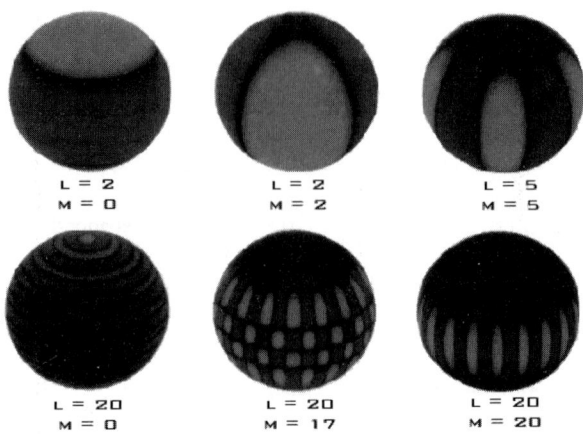

Figure 5.2: Examples of several modes

If P_l^m denotes the associated Legendre function which can be given in an analytical form, the inward or outward motion of points on the surface is related to the value of the real part of the function

$$P_l^m(\cos\Theta)exp(im\phi) \quad (5.14)$$

where Θ, ϕ are spherical polar coordinates. If l, m are low, there is a relatively small number of patches on the solar surface (which oscillate with different directions of radial velocity). If l, m are large, there is a very large number of such patches. We speak of a high degree model if l is large and conversely if l is small. Most of the observable p-modes have periods between 2 and 10 minutes with 5 minutes as a characteristic value. These p-modes are trapped near to the solar surface and in the solar interior. For high values of l the modes are trapped close to the surface. In general the oscillation frequency of any mode depends on the internal properties of the Sun in the region which the mode can propagate. Thus the observation of the oscillations can provide information about the manner in which quantities vary in the solar interior.

If we want to observe solar oscillations we must measure Doppler shifts of the wavelength of solar spectral lines. These shifts are produced by the motion of matter towards or away from the observer.

The $l - \nu$ diagram (Fig. 5.3) is fundamental for helioseismology. This diagram shows how much acoustic energy there is at each frequency for every one of the spatial modes of oscillation. A musical instrument should be tuned to a single frequency and a few harmonious overtones, the Sun resonates in tens of millions of ways all at the same time. The frequency ν of each mode reveals a slightly different part of the Sun's interior. The spatial modes are identified from patterns on the dopplergrams that are made each minute. The frequencies are very low compared to sound waves we are used to hearing. Most of the power is concentrated in a band near 3 mHz, that's one oscillation every 5 minutes (Sound waves we can hear vibrate from tens to thousands of times per second). Higher frequencies aren't trapped inside the Sun, so they don't resonate. Modes with lower ν disappear in the background noise. The spatial scale of the modes is indicated by the angular degree l telling how many node lines there are in the pattern at the surface of the Sun. The l=0 modes are 'breathing' modes where the whole surface of the Sun moves in and out at the same time. Higher order modes divide the surface into a pattern like a checker board, where adjacent squares move in different directions at any given time. A mode of a particular degree, l, at the surface can be associated with resonances having any number of nodes in the radial direction inside the Sun. The number of radial nodes is called the order. The curved lines in the figure are associated with different radial orders. For a given order (line) the frequency decreases with increasing spatial degree. For a give degree, the frequency increases with order. In the Fig. 5.3, the lower left corner is most closely related to what is happening in the core of the Sun. Moving up in frequency or degree tells more about what is happening near the surface. Because sound waves of a particular degree can travel in different directions the lines appear relatively broad. If the material through which any of these modes is travelling is moving, then the measured frequency of the mode is affected. The rotation of the Sun causes the biggest frequency shift and makes the lines shown in the figure broad (frequency shifting). Other motions within the Sun along the path taken by the waves cause different types of frequency changes. Analysis of these frequency changes reveals the internal motions of the Sun.

5.2.1 Theory of Solar Oscillations

Let us briefly describe the basic theory of solar oscillations. We use the basic equations:

$$\rho \frac{d\vec{v}}{dt} = -\text{grad} P + \rho \text{grad} \Phi \tag{5.15}$$

$$\frac{d\rho}{dt} + \rho \text{div} \vec{v} = 0 \tag{5.16}$$

$$\frac{1}{P}\frac{dP}{dt} = \frac{\Gamma}{\rho}\frac{d\rho}{dt} \tag{5.17}$$

$$\nabla^2 \Phi = -4\pi G \rho \tag{5.18}$$

The first equation is the equation of motion, the second the equation of continuity, the third the adiabatic equation and the last is the Poisson equation. Φ denotes the gravitational potential and $\vec{v}$ is the fluid velocity, Γ is an effective ratio of specific heats ($\rho dP/Pd\rho$) which reduces to γ when γ is constant. The time derivative follows the motion of the fluid. It is related to the derivative at a fixed point by $d/dt = \partial/\partial t + \vec{v} \text{grad}$. If we have an equilibrium situation:

$$\rho = \rho_0(r) \quad P = P_0(r) \quad \Phi = \Phi_0(r) \quad \vec{v} = 0 \tag{5.19}$$

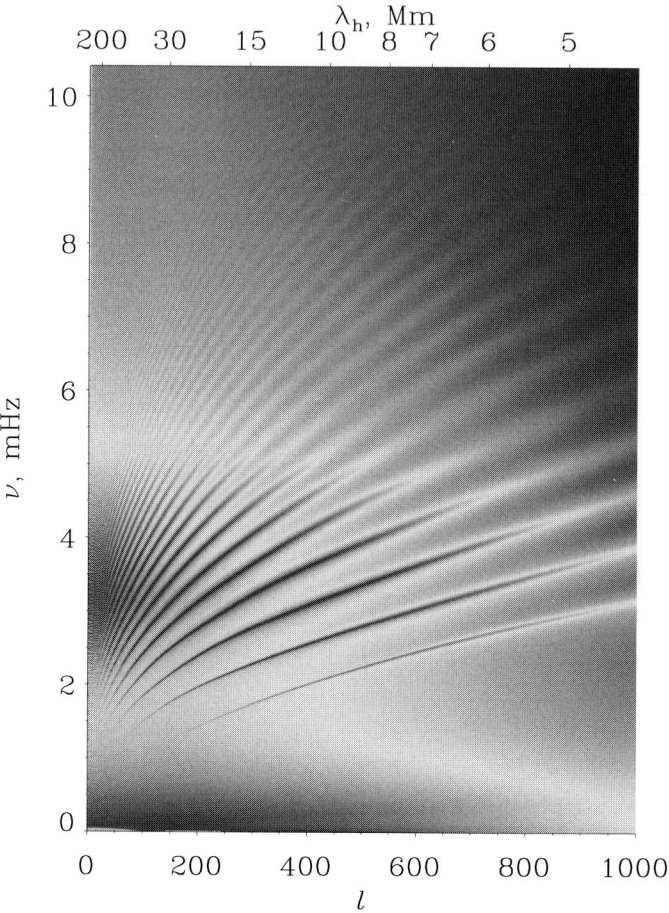

Figure 5.3: l-ν diagram from MDI high-cadence full disk data shows mode frequencies up to 10 mHz and l=1000.

5.2. HELIOSEIMOLOGY-SOLAR OSCILLATIONS

Now we consider small disturbances about this equilibrium in which the perturbed quantities are functions of all the spatial coordinates and the time. In the equilibrium there is no dependence on spherical polar coordinates. For any variable f we can write:

$$f = f_0 + f_1 \qquad f_1 = Re[\exp(i\omega_{nl}t)\bar{f}_1(r)Y_l^m(\Theta, \Phi)] \tag{5.20}$$

The spherical harmonic is given by:

$$Y_l^m(\Theta, \phi) = P_l^m(\Theta)\exp(im\phi) \tag{5.21}$$

If the star is spherical the oscillation frequency does not depend on m. For the Sun, the departure from sphericity is small and the real oscillation modes have a behavior close to that shown above but with different m modes having different frequencies. The oscillation frequency ω depends on n and l. The three numbers n, l, m are related to the numbers of times f_1 vanishes in the radial-, Θ- and ϕ-directions and $m \leq l$.

The functions $f_1(r)$ satisfy a system of differential equations and the boundary conditions have to be defined. Since stars do not have sharp surfaces we may assume to a first approximation that all waves are totally reflected at the surface which is defined as the level where density and pressure vanish. A further simplification arises when the change in the gravitational potential produced by the oscillations is unimportant; for most perturbations this is a good approximation because some parts of the star are moving inwards and others moving outwards. We define a perturbation vector $\vec{\xi}$ by

$$\vec{v} = d\vec{\xi}/dt \tag{5.22}$$

If c_s denotes the velocity of sound in the unperturbed star:

$$c_s = \sqrt{\Gamma P_0/\rho_0} \tag{5.23}$$

one can write:

$$\Psi = c_s^2 \rho_0^{1/2} div\xi \tag{5.24}$$

and the equation for the radial part of ψ is

$$\frac{d^2\psi}{dr^2} = -\frac{1}{c_s^2}\left[\omega^2 - \omega_c^2 - S_l^2\left[1 - \frac{N^2}{\omega^2}\right]\right]\psi \tag{5.25}$$

In addition to the frequency ω we have three frequencies:

- acoustic cut-off frequency ω_c

$$\omega_c^2 = (c_s^2/4H_\rho^2)(1 - 2dH_\rho 2dr) \tag{5.26}$$

Here, $H_\rho = \rho(d\rho/dr)$ denotes the density scale height,

- Lamb frequency S_l

$$S_l = c_s[l(l+1)]^{1/2}/r \tag{5.27}$$

- Brunt-Väissälä frequency N

$$N^2 = g \left[\frac{1}{\Gamma P} \frac{dP}{dr} - \frac{1}{\rho} \frac{d\rho}{dr} \right] \tag{5.28}$$

where $g = GM/r^2$.

S_l is always real but ω_c and N can be imaginary. It can be shown that convection occurs when N^2 is negative. We can write our differential equation for ψ as:

$$\frac{d^2\psi}{dr^2} + K_r^2 \psi = 0 \tag{5.29}$$

We have to consider the sign of K_r^2: for positive K_r^2 we have a sinusodial behavior with radius. For negative K_r^2 we have an exponential dependence giving an exponentially decaying mode which is also called evanescent mode. In reality K_r depends on r and ω. For different values of ω there are regions in the star where the wave propagates and others where it is evanescent. For both, the high frequency range and the low frequency range $4K_r^2$ is positive: for the high frequency range $\omega > S_l, \omega_c$ and pressure fluctuations are most important; we have the p-modes. For low frequencies $\omega < N$ we have the g-modes where the gravity is the restoring force.

As it has been already stated, convection occurs where N becomes imaginary. The p-modes can propagate inside the Sun in a region whose lower boundary is determined by the Lamb frequency and whose upper boundary is given by the acoustic cut-off frequency. The g-modes are trapped beneath the convection zone.

The above given treatment is valid for a spherical non magnetic star. The Sun rotates with a rotation period of approximately 28 days which is small in comparison with the frequencies of the p-modes. It can be shown that for a rotating star the oscillation frequency depends on n and each oscillation frequency ω_{nl} splits into $2l+1$ frequencies ω_{nlm}. Low l modes penetrate close to the center; they can provide some information about the internal rotation of the Sun. High l modes sample the outermost layers.

We only mention here that the study of the effects of the magnetic field on solar oscillations is much more complex.

5.2.2 Helioseismology and Internal Rotation

The rotation of the Sun can be determined quite straightforward: on the one hand tracers such as sunspots or other phenomena visible on the disk can be used, on the other hand, spectroscopic measurements of the plasma can be used. It was found that the Sun does not rotate like a solid body. It rotates faster at the equator (25 days) and slower near the poles (33 days). Moreover, the rotation rate of sunspots at mid-latitudes is somewhat faster than that deduced from Doppler shifts of the surface plasma.

Our Sun is a middle aged star. The surface rotation rates of young solar-type stars are up to 50 times that of the Sun. Our Sun has lost angular momentum through the magnetized solar wind. Therefore, the outer convection zone must

5.2. HELIOSEIMOLOGY-SOLAR OSCILLATIONS

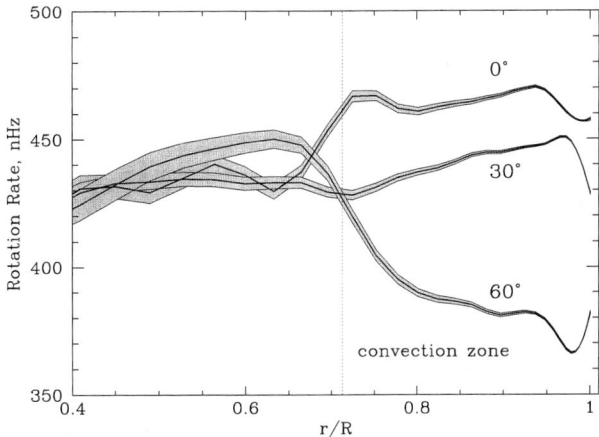

Figure 5.4: This diagram shows the solar rotation rate inferred from two months of MDI Medium-l data as a function of radius at three latitudes, 0 degrees, 30 degrees, and 60 degrees.

have been gradually spinning down. This also had led to the suggestion that the Sun might still posses a rapidly rotating core, perhaps highly magnetized which also could explain the neutrino problem.

It is extremely important to know the internal rotation of the Sun because the interplay between turbulent motions and rotation with magnetic fields is essential for the solar dynamo which leads to the observed 22 year cycles of magnetic activity.

Here, helioseismology can help to understand the internal rotation rate. In a spherically symmetric star the frequencies depend upon n and l but not on m. For each (n, l) pair, there is a $(2l + 1)$ fold degeneracy. Rotation breaks the spherical symmetry and lifts the degeneracy. Advection causes a wave propagation with the Sun's rotation to have a higher measured frequency than a similar wave propagating against rotation. Thus the difference in frequency of a pair of oppositely propagating modes is proportional to m times a weighted average of the rotation rate $\Omega(r, \theta)$ where the modes have appreciable amplitude. Here, $\Omega(r, \theta)$ denotes rotation at radius r and latitude θ. The resulting frequency splitting $\Delta\nu_{nlm}$ is half the value of this difference.

Results on the study of the internal solar rotation rate from the SOHO/MIDI instrument are given in Fig. 5.4.

Measurement of oscillations

How can we measure these oscillations? Let us briefly describe the main principles of a Dopplerimager. Consider the intensity profile of an absorption line. If the material from which this absorption line is emitted moves away from the observer, the line will be redshifted according to the Doppler effect. We can use this effect

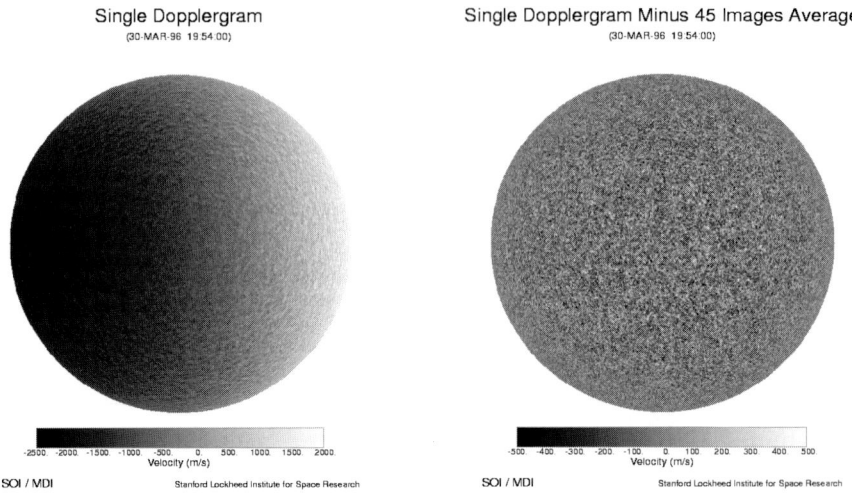

Figure 5.5: MDI Dopplerimage; left: the rotation of the Sun is clearly seen; right: the rotation of the Sun was eliminated and therefore only velocities due to granulation and supergranulation are seen.

to make velocity images of the solar surface. The light from the Sun is sent through a filter that alternates between letting through light from a narrow range of wavelengths on either side of the center of the line. The two light intensities are measured at every point on the solar surface using an imaging camera. The difference between the two intensities changes when the spectral line shifts, and therefore that difference is a measure of the velocity.

Example of Dopplerimages are given in Fig 5.5. In the second example, the average over 45 min was subtracted thus the Doppler effect due to solar rotation is eliminated.

MDI is one of the scientific experiments on SOHO. The medium-l data are spatial averages of the full disk Doppler velocity out to 90% of the solar disk's radius; the measurements are taken every minute and one obtains 23 000 bins of approximately 10 arcsec resolution where solar p modes up to l=300 can be determined. Low-l observables are velocity and continuum intensity images summed into 180 bins where oscillations up to l=20 can be detected. Analyzing such data first asymmetries are detected which arise from the interference between an outward directed wave from the source and a corresponding inward wave that passes through the region of wave propagation. The degree of asymmetry depends on the relative locations of the acoustic sources and the upper reflection layer of the modes. Observations of line profiles of solar modes are therefore suitable to test theories of excitation of solar and stellar oscillations and their interaction with turbulent convection. Concerning the rotation rate, the main result of that inves-

5.2. HELIOSEIMOLOGY-SOLAR OSCILLATIONS

tigation was that:

- Differential rotation: occurs only in the convection zone.

- Radiative interior: rotates almost rigidly.

- Thin shear layer near the surface.

- The transition layer between the radiative and convection zone which is called the *tachocline* is mostly located in the radiative zone and at the equator relatively thin $< 1 R_\odot$ but maybe wider at high latitudes.

- There is a sharp radial gradient of the angular velocity beneath the convection zone and the narrow peak of the sound speed at $0.67\,\mathrm{R_\odot}$ is due to rotationally turbulent mixing in the tachocline.

More details about these results can be found in Kosovichev et al (1998).

Helioseismology can be used also to give arguments in the question of solar neutrinos. Turck - Chièze at al. (2001) used sound-speed and density profiles inferred from SOHO/GOLF and SOHO/ MDI data including these modes, together with recent improvements to stellar model computations, to build a spherically symmetric seismically adjusted model in agreement with the observations. Their model is in hydrostatic and thermal balance and produces the present observed luminosity. In constructing the model, the best physics available was adopted. Some fundamental ingredients were adjusted, well within the commonly estimated errors, such as the p-p reaction rate ($\pm 1\%$) and the heavy-element abundance ($\pm 3.5\%$); the sensitivity of the density profile to the nuclear reaction rates was examined. The corresponding emitted neutrino fluxes demonstrate that it is unlikely that the deficit of the neutrino fluxes measured on Earth can be explained by a spherically symmetric classical model without neutrino flavor transitions.

New insight into the internal structure of the Sun can be obtained by using time-distance helioseimology. Let us explain this technique by considering seismology on earth. Here, the arrival time of the initial onset of a disturbance is measured. If we know the variation of seismic velocity with depth within the earth, then we can calculate the travel time of rays between an earthquake and a receiver using geometrical approximations. So in principle, we can locate any earthquake in both time and space by recording the arrival times of waves at stations worldwide.

In time-distance helioseismology, the travel time of acoustic waves is measured between various points on the solar surface. To some approximation the waves can be considered to follow ray paths; these depend on a mean solar model. The curvature of the ray paths is caused by increasing sound speed with depth below the surface (see Fig. 5.1). The travel time is affected by various inhomogeneities along the ray path, including flows, temperature inhomogeneities and magnetic fields. The technique consists of a measurement of a large number of times between different locations. Then an inversion method is used to construct 3-D maps of the subsurface inhomogeneities. A review article on that technique was given by Duvall et al. (1997).

The Global Oscillation Network Group (GONG) project is a community-based program to conduct a detailed study of solar internal structure and dynamics using helioseismology, by means of observations from a network of six stations spread around the World. To measure solar oscillations one takes a sequence of images of the oscillation pattern at fixed time intervals. The shorter these time intervals between the images, the easier it is to identify the oscillations. Of course the Sun shines only for about half of each day and this causes holes in the time series. In order to avoid these, researchers have established a network of small, partly automated telescopes which are located at a suitable distance in geographic longitude between them. Thus one has at best an uninterrupted view of the Sun. For GONG these stations are Learmont (Australia), Mauna Loa (Hawaii), Big Bear (US, California), Cerro Tololo (Chile), Teide Observatory (Tenerife, Spain), and Udaipur (India). A big problem in such a project is the enormous amount of data. Each station in the network produces more than 200 megabytes of data every day. Details about the instrument used (a Fourier Tachometer) can be found in Beckers et al. (1978).

The BiSON (Birmingham Solar Oscillations Network) project also has six observatories, most of which are automated. As it is explained above the GONG observatories measure the motions on the solar surface caused by the oscillations. The BiSON observatories do so as well, but unlike the GONG network they measure an average velocity over the solar surface (the Sun is observed as as a point source, if it were a star). The measurements therefore are sensitive only to oscillation patterns with very big wavelengths: all smaller-scale patterns are suppressed by being averaged. The two techniques for GONG and BiSON are therefore complementary.

Inversion techniques

As we have explained above, the observed oscillation frequencies depend on the physical structure of the solar interior, e.g the variation of quantities such as ρ, T with r. If we assume a spherical symmetric sun and ignore rotational splitting, then we can deduce from our model of the solar interior the corresponding oscillations. Alternatively one can regard $T, \rho...$ as unknowns and use the observed frequencies in order to obtain them. This is called the inversion method. The total number of quantities that can be determined in such a way is equal to the number of observed oscillations. If more frequencies can be identified, a better model of the internal structure can be obtained.

The Seismic Structure of the Sun from GONG data is described in Gough et al. (1996).

Chapter 6

MHD and the Solar Dynamo

In this chapter we will explain the basic MHD equations which are needed to understand solar active phenomena such as spots, prominences, flares etc. The solar dynamo is needed to maintain the solar activity cycle.

6.1 Solar Magnetohydrodynamics

6.1.1 Basic equations

To understand the surface activity of the Sun and the solar cycle it is necessary to briefly outline the principles of MHD. The properties of electromagnetic fields are described by Maxwell's equations:

$$\nabla \times \vec{H} = \vec{j} + \frac{\partial \vec{D}}{\partial t} \tag{6.1}$$

$$\nabla \times \vec{E} = -\frac{\partial \vec{B}}{\partial t} \tag{6.2}$$

$$\text{div}\,\vec{B} = 0 \tag{6.3}$$

$$\text{div}\,\vec{D} = \rho_E \tag{6.4}$$

Here $\vec{H}, \vec{B}, \vec{D}, \vec{E}, \vec{j}, \rho_E$ are the magnetic field, magnetic induction, electric displacement, electric field, electric current density and electric charge density.

If μ_0, ϵ_0 are the permeability and permittivity of free space, then for most gaseous media in the universe:

$$\vec{B} = \mu_0 \vec{H} \qquad \vec{D} = \epsilon_0 \vec{E} \tag{6.5}$$

The following equation relates the electric current density to the fields producing it (generalized Ohm's law):

$$\vec{j} = \sigma(\vec{E} + \vec{u} \times \vec{B}) \tag{6.6}$$

σ is the electrical conductivity and $\vec{u}$ is the bulk velocity of the matter. The final equations depend on the state of matter; if it consists of electrons and one type of

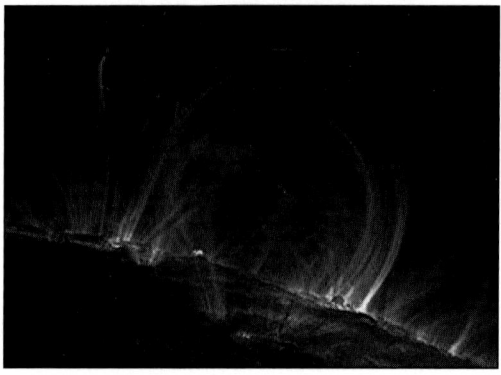

Figure 6.1: Looped magnetic field lines in the solar chromosphere and corona. Photo: NASA

ion:
$$\vec{j} = n_i Z_i e \vec{u}_i - n_e e \vec{u}_e \qquad \rho_E = n_i Z_i e - n_e e \qquad (6.7)$$

$n_i, \vec{u}_i, n_e, \vec{u}_e$ are the number density and velocity of the ions and electrons respectively and $Z_i e, -e$ are the charges on the ion and the electron.

In astrophysics two simplifications are applied:

- magnetic fields are treated as permanent

- electric fields are regarded as transient

The third Maxwell equation (6.3) states that there are no magnetic monopoles. This is a common experience: a division of a permanent magnet into two does not separate north and south poles. Electric fields can be produced by separating positive and negative charges through the fourth Maxwell equation (6.4) however the attraction between these charges is so strong that charge separation is usually cancelled out very quickly. Through the second Maxwell equation electric fields can be produced by time varying magnetic fields. Such fields are only significant, if there are rapid changes by time varying magnetic fields. Magnetic fields produced by the displacement current $\partial \vec{D}/\partial t$ are usually insignificant in astrophysical problems because electric fields are unimportant; however they can be produced by a conduction current $\vec{j}$, if the electrical conductivity is high enough. Such magnetic fields may be slowly variable in time and space.

We therefore neglect $\partial \vec{D}/\partial t$ and combine the equations:

$$\nabla \times \vec{H} = \vec{j} \qquad \nabla \times \vec{E} = -\partial \vec{B}/\partial t \qquad \vec{B} = \mu_0 \vec{H} \qquad \vec{j} = \sigma \vec{E} \qquad (6.8)$$

and obtain:

$$\frac{\partial \vec{B}}{\partial t} + \frac{1}{\mu_0 \sigma} \nabla \times \nabla \times \vec{B} = 0 \qquad (6.9)$$

6.1. SOLAR MAGNETOHYDRODYNAMICS

and using $\nabla \times \nabla \times \vec{B} = \text{grad div}\vec{B} - \nabla^2 \vec{B}$ and $\text{div}\vec{B} = 0$:

$$\frac{\partial \vec{B}}{\partial t} = \frac{1}{\mu_0 \sigma} \nabla^2 \vec{B} \tag{6.10}$$

In cartesian coordinates this equation for the x coordinate is:

$$\frac{\partial \vec{B}_x}{\partial t} = \frac{1}{\mu_0 \sigma} \left[\frac{\partial^2 \vec{B}_x}{\partial x^2} + \frac{\partial^2 \vec{B}_y}{\partial y^2} + \frac{\partial^2 \vec{B}_z}{\partial z^2} \right] \tag{6.11}$$

The solution of these equations shows that magnetic fields decay together with the current producing them. We can derive an approximate decay time: let us assume the currents vary significantly in distance L, then from (6.10) the decay time becomes

$$\tau_D = \mu_0 \sigma L^2 \tag{6.12}$$

If at time $t = 0$ there exists a sinusoidal field

$$B_x = B_0 \exp(iky) \tag{6.13}$$

the solution at a later time t is:

$$B_x = B_0 \exp(iky) \exp(-k^2 t / \mu_0 \sigma) \tag{6.14}$$

The wavelength λ of the spatial variation of the field is $2\pi k$, the original field decays by a factor e in the time $\mu_0 \sigma \lambda^2 / 4\pi^2$.

Let us consider typical fields of stars: the dimension of the star L and the electrical conductivity are both high (if the gas is fully ionized). Therefore, the lifetime of a magnetic field could exceed the main sequence lifetime, such a field is called a *fossil field*.

The same is not true for the Earth. Its field is produced by currents in a liquid conducting core and continuously regenerated by a *dynamo mechanism*.

The electrical conductivity of an ionized gas is $\sim T^{3/2}$. That means that the characteristic time for decay of currents in the outer layers of the Sun is much less than the solar lifetime, whereas the decay near the center exceeds the lifetime (since the temperature near the surface is about 6 000 K and near the center about 1.5×10^7 K). If the field in the solar interior were a fossil field extending throughout the Sun, the field in the outer layers would now be current free - similar to the field of a dipole. However we don't observe this. The surface field is very complex and therefore it must be also regenerated by a dynamo. It is conceivable that a fossil field of the Sun was destroyed at the very early evolution of the Sun, when it was fully convective before reaching the main sequence. Also helioseismology argues against a strong field.

6.1.2 Magnetic Buoyancy

A magnetic field in a conducting fluid exerts a force per unit volume which is

$$\vec{F}_{\text{mag}} = \vec{j} \times \vec{B} = (\text{curl}\vec{B} \times \vec{B})/\mu_0 = -\text{grad}(B^2/2\mu_0) + \vec{B}.\nabla \vec{B}/\mu_0 \tag{6.15}$$

This can be interpreted as:

- grad($B^2/2\mu_0$) isotropic pressure,
- $\vec{B}.\nabla\vec{B}/\mu_0$ tension along the lines of magnetic induction.

The isotropic pressure must be added to the gas pressure: let us assume we have a tube of magnetic flux, and P_{out} denotes the pressure outside and P_{in} the pressure inside the tube:

$$P_{\text{out}} = P_{\text{in}} + B^2/2\mu_0 \tag{6.16}$$

The gas pressure can be written as $\Re\rho T/\mu$, where $\Re$ is the gas constant and μ the mean molecular weight. With $T_{\text{in}} = T_{\text{out}}$ we must have:

$$\rho_{\text{in}} < \rho_{\text{out}} \tag{6.17}$$

A tube of magnetic flux is lighter than its surroundings and will start to rise which is called magnetic buoyancy.

6.1.3 Magnetic Flux Freezing

To a good approximation, the fluid is tied to the magnetic field. Either the fluid motions drag the magnetic field lines around (this happens in the solar photosphere) or the magnetic force is so strong that it constrains the motion of the fluid (this is the case in the outer solar atmosphere). The tying of the fluid to the magnetic field lines also permits the propagation of MHD waves which have some similarity to sound waves but a characteristic speed (Alfvén speed):

$$c_H = \sqrt{B^2/\mu_0\rho} \tag{6.18}$$

The sound speed is given by

$$c_s = \sqrt{\gamma P/\rho} \tag{6.19}$$

6.1.4 The Induction Equation

Let us consider again the Maxwell equations. From $\vec{j} = \sigma(\vec{E} + \vec{u} \times \vec{B})$ we can extract $\vec{E}$:

$$\vec{E} = \frac{\vec{j}}{\sigma} - \vec{u} \times \vec{B} \tag{6.20}$$

This is substituted into the Maxwell equation (6.2) yielding:

$$\nabla \times \left(\frac{\vec{j}}{\sigma} - \vec{u} \times \vec{B}\right) = -\frac{\partial \vec{B}}{\partial t} \tag{6.21}$$

We have already argued that the displacement current can be neglected in the first Maxwell equation and therefore $\nabla \times \vec{B} = \mu\vec{j}$, from which $\vec{j} = 1/\mu\nabla \times \vec{B}$ and

$$\nabla \times \left(\frac{1}{\mu\sigma}\nabla \times \vec{B} - \vec{u} \times \vec{B}\right) = -\frac{\partial \vec{B}}{\partial t} \tag{6.22}$$

6.1. SOLAR MAGNETOHYDRODYNAMICS

using the formula
$$\nabla \times (\nabla \times \vec{A}) = \nabla(\nabla \vec{A}) - \nabla^2 \vec{A} \qquad (6.23)$$
from vectoranalysis, we get:
$$\nabla \times \left(\frac{1}{\mu\sigma} \nabla \times \vec{B}\right) = \frac{1}{\mu\sigma}\left[\nabla(\nabla \vec{B}) - \nabla^2 \vec{B}\right]$$
This gives us the final form of the so called *induction equation*:
$$\frac{\partial \vec{B}}{\partial t} = \nabla \times (\vec{u} \times \vec{B}) + \eta \nabla^2 \vec{B} \qquad (6.24)$$
Here $\eta = 1/\mu\sigma$ is the magnetic diffusivity. The case where the plasma is stationary was already discussed above ($\vec{u} = 0$), the field decays in the ohmic decay time $\tau = L^2/\eta$. Let us discuss the case when $\eta = 0$. Then, the field $\vec{B}$ is completely determined by the plasma motions $\vec{u}$ and the induction equation is the equivalent to the vorticity equation for an inviscid fluid. The magnetic flux Φ through a material surface S which is a surface that moves with the field, is:
$$\Phi = \int_S \vec{B}.d\vec{S} \qquad (6.25)$$
If G is the material closed curve bounding S, the total rate of change of Φ is
$$\frac{D\Phi}{Dt} = \int_S \frac{\partial \vec{B}}{\partial t}.d\vec{S} + \oint_G \vec{B}.(\vec{u} \times dl) \qquad (6.26)$$
$$= \int_S \frac{\partial B}{\partial t}.d\vec{S} + \oint_G (\vec{B} \times \vec{u}).dl \qquad (6.27)$$
$$= \int_S \left[\frac{\partial \vec{B}}{\partial t} - \nabla \times (\vec{u} \times \vec{B})\right].d\vec{S} \qquad (6.28)$$
In the last equation we have used the Stokes Theorem:
$$\int_L (A)\vec{C}d\vec{l} = \int_A (\nabla \times \vec{C})d\vec{A'} \qquad (6.29)$$
If $\eta = 0$ the total flux across any arbitrary surface moving with the fluid remains constant, the magnetic field lines are said to be *frozen* in to the flow.

If v_0, l_0 are typical velocity and length-scale values for our system, then the ratio of the two terms on the right hand side of the induction equation gives the *Magnetic Reynolds Number*
$$R_m = l_0 v_0 / \eta_0 \qquad (6.30)$$
In an active solar surface region one has $\eta_0 = 1 \, \text{m}^{-2}\text{s}^{-1}$, $l_0 = 700 \, \text{km} \sim 1 \, \text{arcsec}$ and $v_0 = 10^4 \, \text{m/s}$ we find $R_m = 7 \times 10^9 \gg 1$. Thus the field is frozen to the plasma and the electric field does not drive the plasma but is simply $\vec{E} = -\vec{u} \times \vec{B}$. However, if the length-scales of the system are reduced the diffusion term $\eta \nabla^2 \vec{B}$ becomes important. Then the field lines are allowed to diffuse through the plasma and this yields to magnetic braking and changing the global topology of the field (magnetic reconnection).

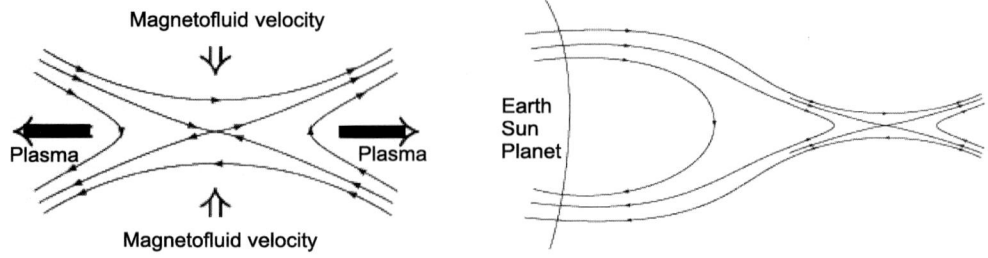

Figure 6.2: Principle of magnetic reconnection

6.1.5 Magnetic Reconnection

Magnetic reconnection is the process by which lines of magnetic force break and rejoin in a lower energy state. The excess energy appears as kinetic energy of the plasma at the point of reconnection. In Fig. 6.2 single arrow lines denote magnetic field and double line arrows the magnetofluid velocity. As it can be seen, the merging of two magnetofluids with oppositely oriented magnetic fields causes the field to annihilate. The excess energy accelerates the plasma out of the reconnection region in the direction of the full double line arrows. Note the characteristic X-point, where the topology changes for the field lines.

The plasma, where the field is annihilated is accelerated outwards to Alfvén speed v_A:

$$v_A = B_0/\sqrt{4\pi M n_B} \tag{6.31}$$

n_B... density inside the current sheet, M the plasma average molecular weight.

A similar process occurs in coronal loops that were observed in hard and soft x-rays by Yohkoh and SOHO instruments. Such a coronal loop (see right drawing in Fig. 6.2) is stretched out by pressure which is provided by buoyancy. A magnetic structure is buoyant because the particle density is lower there since it contains larger magnetic energy density. Thus the external pressure is balanced by a lower gas pressure in conjunction with a magnetic pressure. The top of the loop distends and reconnection occurs. Particles in the reconnection region accelerate towards the surface of the sun and out away. Those particles that are accelerated towards the sun are confined within the loop's magnetic field lines and follow these lines to the footpoint of the loop where they collide with other particles and lose their energy through x-ray emissions. Such processes are the cause of solar flares and will be discussed in the next chapter.

Magnetic reconnection also provides a mechanism for energy to be transported into the solar corona.

A similar process occurs in the earth's magnetotail. The solar wind distends the Earth's dipole field so that the field extends far behind the Earth. Earthward flowing plasma streams with flow velocities up to 1000 km/s (which is close to the local Alfvén speed) have been observed (Birn et al. 1981).

A recent review on solar MHD was given by Walsh (2001)

6.1.6 Fluid Equations

The continuity or mass equation for a fluid is:

$$\frac{D\rho}{Dt} + \rho \nabla . \vec{u} = 0 \qquad (6.32)$$

and the total derivative means here:

$$\frac{D}{Dt} = \frac{\partial}{\partial t} + \vec{u}.\nabla \qquad (6.33)$$

(See any textbook on fluid dynamics for a derivation of this formula). Now let us consider the equation of motion in a plasma with velocity $\vec{u}$: the momentum equation includes the Lorentz force term $\vec{j} \times \vec{B}$ and other forces $\vec{F}$, such as gravity and viscous forces:

$$\rho \frac{D\vec{u}}{Dt} = -\nabla p + \vec{j} \times \vec{B} + \vec{F} \qquad (6.34)$$

Here p is the plasma pressure. Let us assume a Newtonian fluid with isotropic viscosity, then $\vec{F}$ may be written as:

$$\vec{F} = -\rho g(r) \frac{\vec{r}}{r} + \rho \nu \nabla^2 \vec{u} \qquad (6.35)$$

$g(r)$ is the local gravity acting in the radial direction and ν the kinematic viscosity. Let us make thinks more complicated: Consider a frame of reference with angular velocity Ω at a displacement $\vec{r}$ from the rotation axis:

$$\rho \frac{D\vec{u}}{Dt} = -\nabla p + \vec{j} \times \vec{B} + \vec{F} + \rho \left[2\vec{u} \times \vec{\Omega} + \vec{r} \times \frac{d\vec{\Omega}}{dt} + \frac{1}{2}\nabla |\vec{\Omega} \times \vec{r}|^2 \right] \qquad (6.36)$$

The three terms in [ ] denote: Coriolis force, change of rotation and centrifugal force. Stars rotate more rapidly when they are young. Under most circumstances the latter two terms are small compared with the Coriolis term $\vec{u} \times \vec{\Omega}$.

6.1.7 Equation of State

The perfect gas law

$$p = \frac{k\rho T}{m} = nkT \qquad (6.37)$$

determines the constitution of stars. Here m is the mean particle mass and n the number of particles per unit volume. If s denotes the entropy per unit mass of the plasma, then the flux of energy (heat) through a star becomes:

$$\rho T \frac{Ds}{Dt} = -L \qquad (6.38)$$

L is the energy loss function. This function describes the net effect of all the sinks and sources of energy. For MHD applications this becomes:

$$\frac{\rho^\gamma}{\gamma - 1} \frac{D}{Dt} \left(\frac{p}{\rho^\gamma} \right) = -\nabla . \vec{q} + \kappa_r \nabla^2 T + \frac{j^2}{\sigma} + H \qquad (6.39)$$

In this equation we have:

- $\vec{q}$: heat flux due to conduction
- κ_r: coefficient of radiative conductivity
- T temperature
- j^2/σ ohmic dissipation (Joule heating)
- H represents all other sources.

6.1.8 Structured Magnetic Fields

If the plasma velocity is small compared with the sound speed ($\sqrt{\gamma p/\rho}$), the Alfvén speed ($\sqrt{B/\mu\rho}$) and the gravitational free fall speed ($\sqrt{2gl}$), the inertial and viscous terms in equation 6.34 may be neglected yielding:

$$0 = -\nabla p + \vec{j} \times \vec{B} + \vec{F} \tag{6.40}$$

This equation must then be solved with $\nabla \times \vec{B} = ...$, $\nabla \cdot \vec{B} = 0$ and the ideal gas law as well as a simplified form of the energy equation.

Let us introduce the concept of *scale height*. Let

$$0 = -\frac{dp}{dz} - \rho g \tag{6.41}$$

Substitute in the above equation $\rho = pm/kT$ (ideal gas) and integrate:

$$p = p_0 \exp\left[-\int_0^z \frac{dz}{H_p(z)}\right] \tag{6.42}$$

(p_0 is the pressure at $z = 0$). This defines the local pressure scale height H_p:

$$H_P = kT/mg = p/\rho g \tag{6.43}$$

At solar photospheric temperatures ($T \sim 5000$ K) we find $H_p = 150$ km, whereas at coronal temperatures $T \sim 10^6$ K we find $H_p \sim 30$ Mm.

That concept can also be applied to MHD in the case of magnetostatic balance discussed above. Assume that gravity acts along the negative z direction and s measures the distance along the field lines inclined at angle θ to this direction, then the component of 6.40 in the z-direction becomes:

$$0 = -\frac{dp}{ds} - \rho g \cos\theta \qquad dz = ds \cos\theta \tag{6.44}$$

Therefore, the pressure along a given field line decreases with height, the rate of decrease depends on the temperature structure (given by the energy equation).

If the height of a structure is much less than the pressure scale height, gravity may be neglected. The ratio β is given by gas pressure p_0 to magnetic pressure $B_0^2/2\mu$. If $\beta \ll 1$, any pressure gradient is dominated by the Lorentz force and (6.40) reduces to:

$$\vec{j} \times \vec{B} = 0 \tag{6.45}$$

6.1. SOLAR MAGNETOHYDRODYNAMICS

In this case the magnetic field is said to be *force free*. In order to satisfy (6.45) either the current must be parallel to $\vec{B}$ (Beltrami fields) or $\vec{j} = \nabla \times \vec{B} = 0$. In the latter case the field is a current free or potential field.

If β is not negligible and the field is strictly vertical of the form $\vec{B} = B(x)\vec{k}$, then (6.40) becomes:

$$0 = \frac{\partial}{\partial x}\left[p + \frac{B^2}{2\mu}\right] \quad (6.46)$$

6.1.9 Potential Fields

Potential fields result when $\vec{j}$ vanishes. We can write $\vec{B} = \nabla A$ so that $\nabla \times \vec{B} = 0$; with $\nabla \cdot \vec{B} = 0$ one obtains Laplace's equation:

$$\nabla^2 \vec{A} = 0 \quad (6.47)$$

If the normal field component B_n is imposed on the boundary S of a volume V, then the solution within V is unique. Also if B_n is imposed on the boundary S, then the potential field is the one with the minimum magnetic energy.

These two statements have many implications for the dynamics of the solar atmosphere. During a solar flare e.g. the normal field component through the photosphere remains unchanged. However, since enormous amounts of energy are released during the eruptive phase, the magnetic configuration cannot be potential. The excess magnetic energy could arise from a sheared force-free field.

Let us consider an example of a potential field in two dimensions: Consider the solutions $A(x,z) = X(x)Z(z)$ such that $\nabla^2 A = 0$ gives:

$$\frac{1}{X}\frac{d^2 X}{dx^2} = -\frac{1}{Z}\frac{d^2 Z}{Z^2} = -n^2 \quad (6.48)$$

where $n = const$. A solution to (6.48) would be:

$$A = \left(\frac{B_0}{n}\right)\sin(nx)e^{-nz} \quad (6.49)$$

this gives for the field components:

$$B_x = \frac{\partial A}{\partial x} = B_0 \cos(nx)e^{-nz} \quad (6.50)$$

$$B_z = \frac{\partial A}{\partial z} = -B_0 \sin(nx)e^{-nz} \quad (6.51)$$

The result is a two dimensional model of a potential arcade.

6.1.10 3 D Reconstruction of Active Regions

If we look at an active region on the solar disk center we have no information about the 3 D structure of it, especially about the 3 D magnetic field configuration which is important for modelling such regions. Information about the height dependence of active regions can only be obtained when observing such features near the solar limb. Let us consider some simple model to reconstruct these features. Let us assume:

$$B_x = \frac{\partial A}{\partial z}, \; By(x,z), \; B_z = -\frac{\partial A}{\partial x} \quad (6.52)$$

We see immediately that $\nabla \cdot \vec{B} = 0$ Let us assume that the footpoints of the field are anchored down into the photosphere (z=0). Projecting the resulting field onto the xz plane gives:

$$\frac{\partial A}{\partial x}dx + \frac{\partial A}{\partial z}dz = 0 \quad (6.53)$$

Therefore $dA = 0, A = const$. From $\nabla \times \vec{B} = \mu \vec{j}$ calculate the components of the current density:

$$j_x = -\frac{1}{\mu}\frac{\partial B_y}{\partial z} \quad (6.54)$$

$$j_y = \frac{1}{\mu}\left(\frac{\partial B_x}{\partial z} - \frac{\partial B_z}{\partial x}\right) \quad (6.55)$$

$$j_z = \frac{1}{\mu}\frac{\partial B_y}{\partial x} \quad (6.56)$$

And then the components of the Lorentz force ($\vec{j} \times \vec{B}$):

$$\nabla^2 A \frac{\partial A}{\partial x} + B_y \frac{\partial B_y}{\partial x} = 0 \quad (6.57)$$

$$\frac{\partial B_y}{\partial x}\frac{\partial A}{\partial z} - \frac{\partial A}{\partial x}\frac{\partial B_y}{\partial z} = 0 \quad (6.58)$$

$$\nabla^2 A \frac{\partial A}{\partial z} + B_y \frac{\partial B_y}{\partial z} = 0 \quad (6.59)$$

6.1.11 Charged Particles in Magnetic Fields

In this chapter we consider first the motion of a single charged particle in a given electromagnetic field. The particle has charge q and the equation of motion is:

$$m\frac{d\vec{u}}{dt} = q(\vec{E} + \vec{u} \times \vec{B}) \quad (6.60)$$

Let us write:

$$\vec{B_0} = B_0 \vec{b}$$
$$\vec{E_0} = E_\| \vec{b} + \vec{E}_\perp$$
$$\vec{v} = v_\| \vec{b} + \vec{v}_\perp$$

where $\vec{b}$ is a unit vector and $\vec{v} \times \vec{B_0} = B_0(\vec{v}_\perp \times \vec{b})$ is perpendicular to $\vec{b}$ equation 6.60 splits into a parallel and a perpendicular component:

$$m\frac{dv_\|}{dt} = qE_\| \quad (6.61)$$

$$m\frac{d\vec{v}_\perp}{dt} = q[\vec{E}_\perp + B_0(\vec{v}_\perp \times \vec{b})] \quad (6.62)$$

6.1. SOLAR MAGNETOHYDRODYNAMICS

Equation 6.61 has the solution

$$v_{\|} = (qE_{\|}/m)t + v_{\|0} \tag{6.63}$$

Here $v_{\|0}$ is the velocity component at $t = 0$ in the direction of the magnetic field line. We see that particles of opposite sign of charge move in opposite directions, they move along an electric field parallel to a magnetic field which destroys $E_{\|}$.

Now let us solve equation 6.62 by writing:

$$\vec{v}_{\perp} = \vec{v}'_{\perp} + \vec{E}_{\perp} \times \vec{b}/B_0 \tag{6.64}$$

and equation 6.62 becomes:

$$m\frac{d\vec{v}'_{\perp}}{dt} = qB_0(\vec{v}'_{\perp} \times \vec{b}) \tag{6.65}$$

Summarizing we arrive at:

- Motion in which acceleration is $\perp$ to the velocity,
- const. acceleration to the velocity.

This is a motion in a circle around the direction of $\vec{b}$ and the motion has frequency $|q|B_0/m$, the magnitude of the velocity is $v_{\perp 0}$; the radius of the orbit, the gyration radius r_g is

$$r_g = mv_{\perp 0}/|q|B \tag{6.66}$$

For an electron the gyration frequency is 1.8×10^{11}(B/Tesla)Hz. The corresponding gyration radius is $6 \times 10^{-9}(v_{\perp 0}/\text{km/s})(B/\text{T})$m. Summarizing the motion of a particle:

- accelerated motion along the field lines,
- circular motion around the field,
- drift velocity $\vec{E}_{\perp} \times \vec{b}/B_0$ perpendicular to both electric and magnetic fields,
- the sense of the accelerated the circular motions depends on the sign of the electric charge.
- the drift velocity is the same for all particles,
- in the absence of electric fields, a particle moves with a constant velocity in the direction of the magnetic field and with a velocity of constant magnitude around the field, thus it moves along a helical path. In all this discussion we have neglected one important effect. Accelerated charged particles radiate, for non relativistically moving particles this radiation is known as *cyclotron radiation* and for relativistic particles as *synchrotron radiation*.

Finally, if there is a constant non magnetic force $\vec{F}$ perpendicular to $\vec{B}$, there is a drift velocity:

$$v_{\mathrm{DF}} = \vec{F} \times \vec{B}/qB^2 \quad (6.67)$$

Please note that again $\vec{v}_{\mathrm{DF}}$ is charge dependent. Let us assume that $\vec{F}$ is the gravitational force $\vec{F} = m\vec{g}$, then

$$\vec{v}_{\mathrm{DF}} = m\vec{g} \times \vec{B}/qB^2 \quad (6.68)$$

Thus the drift velocity depends on the mass/charge ratio, the ion drift is much larger than the electron drift; the particles drift in opposite directions, a current is produced.

Let us consider a large assembly of particles; these particles interact which is called collision. If τ_c is the characteristic time between collisions the collision frequency is $\nu_c = 1/\tau_c$. If ν_c is large, the particle motions will be disordered and decoupled from the magnetic field, the fluid will not be tied to the field. If collisions are relatively rare, not only individual particles but the whole fluid will be tied to the field. The collisions provide the electrical resistivity of matter; in a fully ionized gas a good approximation to the value of the electrical conductivity is:

$$\sigma = n_e e^2 \tau_c / m_e \quad (6.69)$$

and $\tau_c \sim T^{3/2}$.

6.1.12 MHD Waves

The equation that describes the connections between the force exerted by the magnetic field and the fluid motions is

$$\rho \frac{d\vec{v}}{dt} = -\mathrm{grad} P + \vec{j} \times \vec{B} + \rho \mathrm{grad}\phi \quad (6.70)$$

The forces on the gas are the gas pressure P, the gravitational potential ϕ and the magnetic force $\vec{j} \times \vec{B}$. For a full description of the system we write down two additional equations:

a) equation of continuity (conservation of mass):

$$\frac{d\rho}{dt} + \rho \mathrm{div}\vec{v} = 0 \quad (6.71)$$

b) The relation between P and ρ e.g. in the adiabatic form

$$\frac{1}{P}\frac{dP}{dt} = \frac{\gamma}{\rho}\frac{d\rho 0}{dt} \quad (6.72)$$

Note that d/dt is the rate of change with time following a fluid element moving with velocity $\vec{v}$:

$$\frac{d}{dt} = \frac{\partial}{\partial t} + \vec{v}.\mathrm{grad} \quad (6.73)$$

6.1. SOLAR MAGNETOHYDRODYNAMICS

Consider the simplest case: a medium with uniform density ρ_0, pressure P_0, containing a uniform magnetic field $\vec{B}_0$. We ignore the influence of the gravitational field and assume that σ is so large that $\vec{E} + \vec{v} \times \vec{B} = 0$. Now let us assume a perturbation for any variable in the form of:

$$f_1 \sim \exp i(\vec{k}.\vec{r} - \omega t) = \exp i(k_x x + k_y y + k_z z - \omega t) \quad (6.74)$$

$\vec{k}$ is the wave vector, ω the wave frequency. The dispersion relation between ω and $\vec{k}$ when there is no magnetic field is:

$$\omega^2 = k^2 c_s^2 \quad (6.75)$$

Therefore, in that case only one type of waves can propagate – sound waves. The wave propagates through the fluid at the wave speed $c_s = \omega/k$, $k = |\vec{k}|$, which is called the phase velocity of the wave.

If there is a magnetic field, the force $\vec{j} \times \vec{B}$ couples to the equation and also the Maxwell equations must be taken into account. It is very important to note that the magnetic field introduces a preferred direction into the system. In a uniform medium, sound waves travel equally strongly in all directions from its source, this is not true for MHD waves. If we write the magnetic field again in the form $\vec{B}_0 = B_0 \vec{b}$, then we find three types of MHD waves:

- Alfvén waves : the dispersion relation is given by

$$\omega^2 = (\vec{k}.\vec{b})^2 c_H^2 \quad (6.76)$$

- fast and slow magnetosonic waves; their dispersion relation is given by:

$$\omega^4 - \omega^2 k^2 (c_s^2 + c_H^2) + k^2 (\vec{k}.\vec{b})^2 c_s^2 c_H^2 = 0 \quad (6.77)$$

Let us consider two special cases: if the waves propagate along the field $\vec{k}.\vec{b} = k$, there are two waves with $\omega^2 = k^2 c_H^2$ and the sound wave $\omega^2 = k^2 c_s^2$ unaffected by the field. For wave propagation perpendicular to the field only one wave survives with $\omega^2 = k^2(c_s^2 + c_H^2)$. When waves propagate anisotropically, it is necessary to introduce another wave velocity in addition to the phase velocity, the group velocity, given by $\partial \omega / \partial \vec{k}$ with which the wave carries energy or information. The group velocity for Alfvén waves is always $c_H \vec{b}$. What does that mean? Regardless of the direction in which it propagates, energy always travels along the field lines with speed c_H.

6.1.13 Magnetic Fields and Convection

Let L be the length of a box, v a typical velocity. The magnetic diffusivity is $1/\mu_0 \sigma$, the eddy turnover time L/v. The resistive decay time is then $\mu_0 \sigma L^2$ and the resistive decay time/eddy turnover time is denoted as magnetic Reynolds number for the flow.

$$R_m = L v \mu_0 \sigma \quad (6.78)$$

If the magnetic Reynolds number R_m is very low, the field is unaffected by the motions, if it is high, it is wound up many times before dissipation occurs. For an intermediate value of R_m the magnetic field is carried from the center of the eddy becoming concentrated in flux ropes at the edge. This buoyant flux ropes rise towards the surface and this leads to the appearance of sunspots. However we must also take into account that convection involves different length scales. Large eddies affect the overall structure of the magnetic field as it has been just described. Others may be influenced e.g. granulation. Granulation is suppressed in a sunspot. As it was shown earlier, in the absence of magnetic field convection occurs in a gas, if the ration of the temperature gradient to the pressure gradient satisfies the relation:

$$\frac{P}{T}\frac{\mathrm{d}T}{\mathrm{d}P} > \frac{\gamma - 1}{\gamma} \tag{6.79}$$

If a vertical magnetic field of strength B threads the fluid, then this has to be modified to:

$$\frac{P}{T}\frac{\mathrm{d}T}{\mathrm{d}P} > \frac{\gamma - 1}{\gamma} + \frac{B^2}{B^2 + \gamma P} \tag{6.80}$$

Thus a strong magnetic field can prevent convection and a weaker field can interfere with convection. Note also that the magnetic field cannot prevent motions which are oscillatory up and down the field lines hut these are likely to be less efficient at carrying energy.

6.2 The Solar Dynamo

So far we have discussed the different aspects of solar activity. In the section on MHD it was shown that due to dissipation, such recurrent phenomena on the solar surface and atmosphere cannot be explained by just assuming a fossil magnetic field of the Sun. Therefore, many attempts had been made in order to explain the recurrent solar activity phenomena such as sunspots, their migration toward the equator in the course of an activity cycle etc. In the first section of this paragraph we will give a general description of the basic dynamo mechanism, in the following chapter some formulas are given.

6.2.1 Basic Dynamo Mechanism

Let us briefly recall what are the observational facts that a successful model for the solar dynamo must explain:

- 11 year period of the sunspot cycle; not only the number of sunspots varies over that period but also other phenomena such as the occurrence of flares, prominences,.... etc.

- the equator-ward drift of active latitudes which is known as Spörers law and can be best seen in the butterfly diagram. At the beginning of a cycle active regions appear at high latitudes and toward the end they occur near the equator.

6.2. THE SOLAR DYNAMO

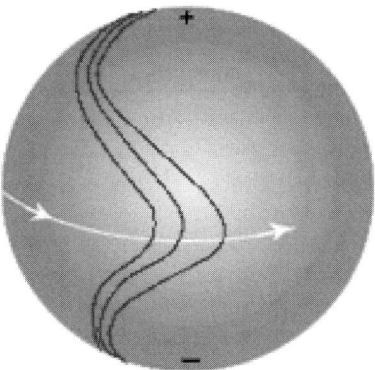

Figure 6.3: Illustration of the ω effect. The field lines are wraped around because of the differential rotation of the Sun

- Hale's law: as we have mentioned the leader and the follower spot have opposite polarities. This reverses after 11 years for each hemisphere so that the magnetic cycle is in fact 22 years.

- Sunspot groups have a tilt towards the equator (this is sometimes also called Joy's law).

- Reversal of the polar magnetic fields near the time of the cycle maximum.

As we know from fundamental physics, magnetic fields are produced by electric currents. How are these currents generated in the Sun? The solar plasma is ionized and it is not at rest. There are flows on the solar surface as well as in the solar interior producing magnetic fields which contribute to the solar dynamo.

The ω Effect

Let us consider magnetic fields inside the Sun. There the conditions require that the field lines are driven by the motion of the plasma. Therefore, magnetic fields within the Sun are stretched out and wound around the Sun by differential rotation (the Sun rotates faster at the equator than near the poles). Let us consider a north-south orientated magnetic field line. Such a field line will be wraped once around the Sun in about 8 months because of the Sun's differential rotation (Fig. 6.3).

The α Effect

However, the field lines are not only wraped around the Sun but also twisted by the Sun's rotation. This effect is caused by the coriolis force. Because the field lines become twisted loops, this effect was called α effect. Early models of the dynamo assumed that the twisting is produced by the effects of the Sun's rotation on very large convective flows that transport heat to the Sun's surface. The main problem of that assumption was, that the expected twisting is too much and would

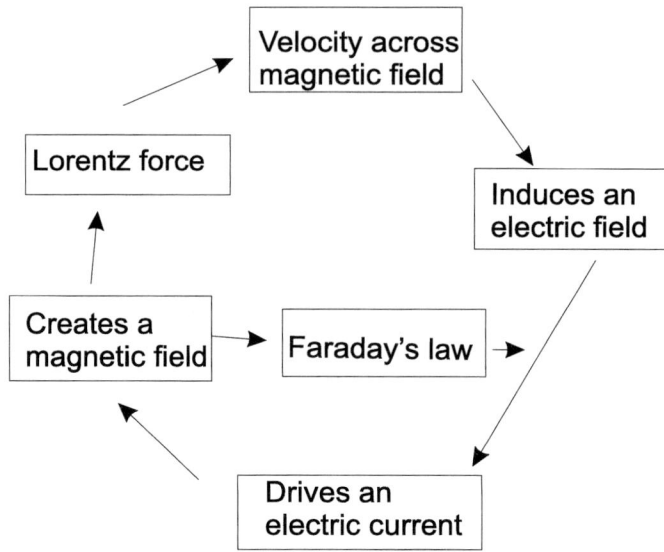

Figure 6.4: The MHD relation between flows and magnetic fields

produce magnetic cycles of only a couple of years. More recent dynamo models assume that the twisting is due to the effect of the Sun's rotation on rising flux tubes. These flux tubes are produced deep within the Sun.

The interface between radiation zone and convection zone

If dynamo activity occurs throughout the entire convection zone the magnetic fields within that zone would rapidly rise to the surface and would not have enough time to experience either the alpha or the omega effect. This can be explained as follows: a magnetic field exerts a pressure on its surroundings ($\sim B^2$, proportional to its strength). Therefore, regions of magnetic fields will push aside the surrounding gas. This produces a bubble that rises continuously to the surface. However such a buoyancy is not produced in the radiation zone below the convection zone. Here, the magnetic bubble would rise only a short distance before it would find itself as dense as its surroundings. Consequently, it is assumed that magnetic fields are produced at this interface layer between the radiation zone and the convection zone.

The Meridional Flow

The solar meridional flow is a flow of material along meridional lines from the equator toward the poles at the surface and from the poles to the equator deep inside. At the surface this flow is in the order of 20 m/s, but the return flow toward the equator deep inside the Sun must be much slower since the density is

6.2. THE SOLAR DYNAMO

much higher there- maybe between 1 and 2 m/s. This slow plasma flow carries material from the polar region to the equator in about 20 years.

Thus the energy that drives the solar dynamo comes from a) rotational kinetic energy, b) another part in the form of small-scale, turbulent fluid motions, pervading the outer 30% in radius of the solar interior (the convection zone).

Mathematical description

Let us discuss some basic mathematics. In the magnetohydrodynamic limit the dynamo process is described by the induction equation:

$$\frac{\partial \vec{B}}{\partial t} = \nabla \times (\vec{u} \times \vec{B}) - \nabla \times (\eta_e \nabla \times \vec{B}) \tag{6.81}$$

The flow $\vec{u}$ is a turbulent flow. In the mean-field electrodynamics one makes the following assumptions: magnetic and flow fields are expressed in terms of a large-scale mean component and a small scale fluctuating (turbulent) component. If we average over a suitably chosen scale we obtain an equation that governs the evolution of the mean field. This is identical to the original induction equation but there appears a mean electromotive force term associated with the (averaged) correlation between the fluctuation velocity and magnetic field components. The basic principles of mean field electrodynamics were given by Krause and Rädler(1980). The velocity and the field are expressed as:

$$\vec{u} = <\vec{u}> + \vec{u}' \qquad \vec{B} = <\vec{B}> + \vec{B}' \tag{6.82}$$

$<\vec{u}>, <\vec{B}>$ represent slowly varying mean components and $\vec{u}', \vec{B}'$ non axisymmetric fluctuating components. The turbulent motion $\vec{u}'$ is assumed to have a correlation time τ and a correlation length λ which are small compared to the scale time t_0 and scale length l_0 of the variations of $<\vec{u}>$ and $<\vec{B}>$. In other words, τ is a mean time after which the correlation between $\vec{u}'(t=\tau)$ and $\vec{u}'(t=0)$ is zero and λ is comparable to the mean eddy size. We assume that $<\vec{u}'>, <\vec{B}'> = 0$.

If this is substituted into the induction equation:

$$\frac{\partial <\vec{B}>}{\partial t} = \nabla \times (\vec{E} + <\vec{u}> \times <\vec{B}>) - \nabla \times (\eta \nabla \times <\vec{B}>) \tag{6.83}$$

This is subtracted from the complete equation:

$$\frac{\partial \vec{B}'}{\partial t} = \nabla \times (<\vec{u}> \times \vec{B}' + \vec{u}' \times <\vec{B}> + \vec{G}) - \nabla \times (\eta \nabla \times \vec{B}') \tag{6.84}$$

where

$$\vec{E} = <\vec{u}' \times \vec{B}'> \qquad \vec{G} = \vec{u}' \times \vec{B}' - <\vec{u}' \times \vec{B}'> \tag{6.85}$$

$\vec{E}$ is a mean electric field that arises from the interaction of the turbulent motion and the magnetic field. This field must be determined by solving the equation for $\vec{B}'$ and here several assumptions are made. First of all we stressed that $<\vec{v}'> = 0$.

This may be a good assumption when considering a fully turbulent velocity field. However in the Sun we are dealing with a sufficiently ordered convective field where the Coriolis force plays an important role. The other approximation is a first order smoothing: $\vec{G} \sim 0$. That is valid only if $\vec{B}' << <\vec{B}>$. Then our equation reduces to:

$$\frac{\partial \vec{B}'}{\partial t} + \nabla \times (\eta \nabla \times \vec{B}') = \nabla \times (\vec{u}' \times <\vec{B}>) \tag{6.86}$$

We want to determine $\vec{E}$. Thus only $\vec{B}''$ the component of $\vec{B}'$ which is correlated with $\vec{u}'$ must be considered. By definition τ, $\vec{B}(t+\tau)$ is not correlated with $\vec{B}(t)$ for any t. $\vec{B}''(t)$ may be determined by integration of the above equation from $t - \tau$ to t. Note also, that the order of the convective turn over time $\tau \sim \lambda/v$ and thus both $\vec{u}'$ and $<\vec{B}>$ may be regarded as independent of t. Thus the integration yields:

$$E_i = \alpha_{ij} <B_{ij}> + \beta_{ijk} \frac{\partial <B_j>}{\partial x_k} \tag{6.87}$$

where α_{ij}, β_{ijk} depend on the local structure of the velocity field and on τ. If the turbulent field is isotropic, then $\alpha_{ij} = \alpha \delta_{ij}$, $\beta_{ij} = \beta \epsilon_{ijk}$, and

$$\vec{E} = \alpha <\vec{B}> - \beta \nabla \times <\vec{B}> \tag{6.88}$$

If τ is small compared to the decay time λ^2/η, the diffusive term may be neglected and from 6.86 we get

$$\alpha = -\frac{1}{3}\tau <\vec{u}'.\nabla \times \vec{u}'>, \qquad \beta = \frac{1}{3}\tau v^2 \tag{6.89}$$

And finally:

$$\frac{\partial \vec{B}}{\partial t} = \nabla \times (\alpha \vec{B} + \vec{u} \times \vec{B}) - \nabla \times [(\eta + \beta)\nabla \times \vec{B}] \tag{6.90}$$

Compared to the normal induction equation, this contains the term $\alpha \vec{B}$ and the eddy-diffusivity coefficient β. In the mean field dynamo, the magnetic diffusivity η is replaced by a total diffusivity $\eta' = \eta + \beta$ and the equation becomes:

$$\frac{\partial B}{\partial t} = \nabla \times (\alpha \vec{B} + \vec{u} \times \vec{B}) + \eta' \nabla^2 \vec{B} \tag{6.91}$$

Please note that most often the prime is dropped on η; however, in the presence of α it is implied to use the turbulent diffusivity. It is assumed that $\vec{B}$ is axisymmetric. Then it can be represented by its poloidal and toroidal components $A(x, z, t)$ and $B(x, z, t)$ and $\vec{u} = u(x, z, t)\vec{j}$. Neglecting the advection terms:

$$\left(\frac{\partial}{\partial t} - \eta \nabla^2\right) B = [\nabla u \times \nabla A].\vec{j} - \alpha \nabla^2 A \tag{6.92}$$

$$\left(\frac{\partial}{\partial t} - \eta \nabla^2\right) A = \alpha B \tag{6.93}$$

6.2. THE SOLAR DYNAMO

Note that the dynamo action is possible because we have a regeneration of both toroidal and poloidal fields. Let us consider the source term in the first of the two above equations. ∇u describes a non uniform rotation. It can be argued that this term is larger than the next term involving α. This set of equations then describes the so called $\alpha - \omega$-dynamo. The equations describe:

- ω effect: the poloidal field is sheared by non uniform rotation to generate the toroidal field.

- α effect: this is the essential feedback. The helicity $\vec{v}_c \cdot \nabla \times \vec{v}_c$ of the non axisymmetric cyclonic convection generates an azimuthal electromotive force $\vec{E}$ which is proportional to the helicity and to B_ϕ.

Let us define a characteristic length scale l_0, a decay time $t_0 = l_0^2/\eta$ and $u = s_0 \omega$, where s_0 is of the order of the local radius of rotation and ω the local angular velocity. We may rewrite the above equations in terms of the non dimensional variables $t' = t/t_0$ and $\vec{r}' = \vec{r}/l_0$. By an elimination of B and neglection of the α^2 terms we arrive at

$$\left(\frac{\partial}{\partial t'} - \nabla'^2\right)^2 A = \frac{\alpha l_0^2 s_0}{\eta^2}[\nabla'\omega \times \nabla' A].\vec{j} \qquad (6.94)$$

If α_0 and ω_0 are scale factors giving the orders of magnitudes of α and $|\nabla'\omega|$ then

$$\left(\frac{\partial}{\partial t'} - \nabla'^2\right)^2 A = D\frac{\alpha}{\alpha_0}\left[\frac{\nabla'\omega}{\omega_0} \times \nabla' A\right].\vec{j} \qquad (6.95)$$

In that equation the non dimensional *dynamo number D* is

$$D = \frac{\alpha \omega_0 l_0^2 s_0}{2\eta^2} \qquad (6.96)$$

It is extremely important to note that the onset of a dynamo action depends on D. If D for a given system exceeds some critical value than there will be dynamo action. Examining our set of equations we may also note that dynamo action is possible when ∇u is negligible compared to α. Such dynamos are called α^2 dynamos. If both terms of the source term are comparable then we speak of an $\alpha^2 \omega$ dynamo.

Solar like stars have well developed and structured convection zones. Thus, the $\alpha - \omega$ dynamo is the most likely dynamo mode.

We have already discussed the results of helioseismic observations which have shown that the observed surface latitudinal differential rotation persists to the base of the convection zone. Below this zone the angular velocity rapidly changes to solid body rotation at a rate equal to the surface mid-latitude. Therefore, three distinct dynamo modes can exist:

- polar interface mode: the alpha effect is concentrated near the poles. The toroidal field regeneration occurs through the agency of the negative radial shear at high latitudes below the core-envelope interface.

- equatorial interface mode: the dynamo mode is concentrated at low latitudes; here the regeneration of the toroidal field occurs through the agency of the positive radial shear at low latitudes, below the core-envelope interface.

- hybrid mode: this dynamo mode covers all latitudes, the toroidal field regeneration occurs through the agency of the positive latitudinal shear which exists at all latitudes above and below the core-envelope interface. It requires shear both above and below the interface.

Finally let us discuss some nonlinear effects. As long as the Lorentz force associated with the growing magnetic fields remains small enough not to impede the driving flows the above described model remains valid. One can also say that as long as the magnetic energy remains smaller than the kinetic energy of the fluid motions the theory is valid. If the two energies are the same, than we speak of equipartition. At the base of the solar convection zone, the equipartition field is about 10 000 Gauss. When this is reached, there is a backreaction of the magnetic field on the flow. This causes a stop of the dynamo's growth. Unfortunately recent numerical simulations have suggested that dynamo action actually ceases long before equipartition with the mean field is attained. Thus a turbulent hydromagnetic dynamo cannot produce structured, large-scale mean magnetic field of strength significant with respect to equipartition. But of course the Sun produces them!

The backreaction of the dynamo-generated field on the small scale turbulent motions can be introduced by a modification of the α:

$$\alpha \to \alpha(<\vec{B}>) = \frac{\alpha_0}{1 + \left(\frac{|<\vec{B}>|}{B_{eq}}\right)^2} \tag{6.97}$$

This is called α quenching. The equation tells us, that once the mean magnetic field reaches B_{eq} (which is the value for equipartition), the alpha effect is suppressed. Recently it was also proposed to take into account the extremely high Reynolds numbers R_m and to describe the alpha quenching by:

$$\alpha \to \alpha(<\vec{B}>) = \frac{\alpha_0}{1 + R_m \left(\frac{|<\vec{B}>|}{B_{eq}}\right)^2} \tag{6.98}$$

This takes into account that the small-scale component of the dynamo generated magnetic field reaches equipartition with small scale turbulent fluid motions long before the mean field does. Reviews on the solar dynamo and the emergence of magnetic flux at the surface can be found in Fisher et al. (2000) and Moreno-Insertis (1994).

So far we have discussed large dynamos which are invoked to explain the origin of the solar cycle and of the large scale component of the solar magnetic field. We should add here that the origin of small scale magnetic fields can also be understood in terms of dynamo processes. Recent advances in the theory of dynamo operating in fluids with high electrical conductivity – fast dynamos, indicate that most sufficiently complicated chaotic flows should act as dynamos (Cattaneo, 1999). The existence of a large scale dynamo is related to the breaking of symmetries in the underlying field of turbulence (Cattaneo, 1997).

6.3. SOLAR ACTIVITY PREDICTION

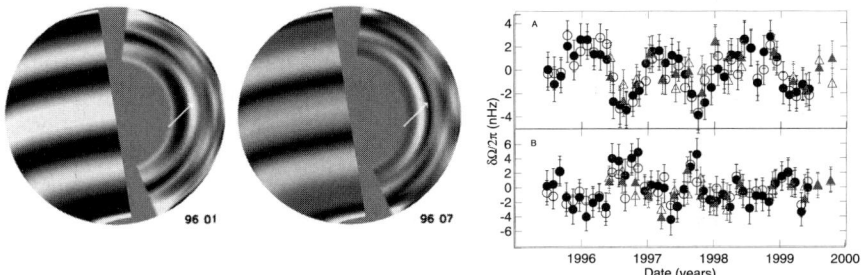

Figure 6.5: a) Cutaway images of solar rotation showing a peak and a trough of the 0.72R variation, with black indicating slow rotation, grey intermediate, and white fast. b) Variations with time of the difference of the rotation rate from the temporal mean at two radii deep within the Sun, with the site at 0.72 $R_\odot$ located above the tachocline and that at 0.63 $R_\odot$ below it, both sampling speeding up and slowing down in the equatorial region. Results obtained from GONG data for two different inversions are shown with black symbols, those from MDI with red symbols. (Image courtesy NSF's National Solar Observatory)

Observations form SOHO

Near the base of the convection zone the analysis of solar oscillations (data from the SOHO/MDI) has shown that there exist variations in the rotation rate of the Sun. A successive acceleration and deceleration with a strange period of 1.3 years was found near the equator and 1.0 years at high latitudes. The largest temporal changes were found both above and below the 'tachocline', a layer of intense rotational shear at the interface between the convection zone and the radiation zone. The variations near the equator are strikingly out of phase above and below the tachocline, and involve changes in rotation rate of about 6 nHz, which is a substantial fraction of the 30 nHz difference in angular velocity with radius across the tachocline. The solar magnetic dynamo is thought to operate within the tachocline, with the differential rotation there having a crucial role in generating the strong magnetic fields involved in the cycles of solar activity. This is illustrated in Fig. 6.5.

6.3 Solar Activity Prediction

Generally, prediction of solar activity is related to the problem of prediction of a given time series since solar activity parameters such as sunspot numbers are given as a function of time. Therefore, the problem can be examined on the basis of recent nonlinear dynamics theories. The solar cycle is very difficult to predict due to the intrinsic complexity of the related time behavior and to the lack of a successful quantitative theoretical model of the Sun's magnetic cycle. Sello (2001) checked the reliability and accuracy of a forecasting model based on concepts of nonlinear dynamical systems applied to experimental time series, such as embedding phase space, Lyapunov spectrum, chaotic behavior. The model is based on a local hypothesis of the behavior on embedding space, utilizing an optimal number of neighbor vectors to predict the future evolution. The main task is to set up and to compare a promising numerical nonlinear prediction technique, essentially based on an inverse problem, with the most accurate prediction methods, like the so-called "precursor methods" which appear now reasonably accurate in predicting "long-term" Sun activity, with particular reference to "solar" and "geomagnetic" precursor methods based on a solar dynamo theory.

Snodgrass (2001) studied azimuthal wind bands known as the *torsional oscillations*. These have been revealed primarily by studying the longitudinally averaged solar rotation over a period spanning several full solar rotations. This averaging yields what look like broad but slow, oppositely-moving (∼5 m/s) bands lying to either side of the centroid of the sunspot butterfly, making the activity band appear to be a zone of weakly enhanced shear. The torsional pattern tells us something about the cycle, and since it precedes the onset of activity, it might be useful as a predictor of the level of activity to come. For the present cycle 23, the torsional pattern did not emerge until just before solar minimum, whereas for cycles 21 and 22 it appeared several years earlier. This would have suggested by 1996 the cycle 23 would be weaker than the previous two.

Calvo et al. (1995) used the neural network technique to analyze the time series of solar activity (given by the Wolf number).

Hernandez (1993) also used neural nets to construct nonlinear models to forecast the AL index given solar wind and interplanetary magnetic field (IMF) data.

Gleisner and Lundstedt (2001) used a neural network-based model for prediction of local geomagnetic disturbances. Boberg et al. (2000) made real time Kp predictions from solar wind data using neural networks.

6.4 Stellar Activity

The Sun is the only star that permits a two-dimensional study of its activity. However, it is only a single set of stellar parameters, since its mass, composition and evolutionary status are fixed. Stars are one-dimensional objects when observed from the Earth but they cover a wide range of physical parameters. Thus the solar-stellar connection is essential for a better understanding of solar phenomena as well as for stellar phenomena. In the 40s e.g. the solar chromosphere was thought to be unique.

The first detection of stellar activity phenomena were made by the observation of magnetic fields. Field strengths in the range of 1-2 kGauss can only be measured by a comparison of magnetically sensitive lines with magnetically insensitive lines. It was surprising that these stars seem to be covered by such strong fields about 20-80% of the total surface (the Sun is only covered ≤1%). The problem is, that by these methods coverages lower than 20% cannot be detected. That means that the Sun's magnetic field would not have been detected if it were at the distance of these stars. More than 100 years ago Pickering suggested that luminosity fluctuations in stars of the order of 20% over periods of days or a few weeks might indicate that they are spotted. In the 1970 extensive investigations were performed to

6.4. STELLAR ACTIVITY

explain luminosity variations of e.g. the RSCVn stars or BY Draconis stars (having luminosities $< 1/2 L_\odot$). The observed lightcurves required circular spots. The RSCVn stars occur in binary stars were tidal interactions play an important role, therefore their starspots are quite different from the sunspots. BY Dra stars are rapidly rotating young low massive stars characterized by intense chromospheric emission. Large spots on the Sun cause a variation of the integrated flux $< 1\%$, whereas up to 30 % for RSCVn and BY Dra stars.

Also flares were detected on stars. Here it is extremely important to have observations in the EUV/X ray window. Generally pre main sequence stars show high levels of magnetic activity and strong flares. FU Orionis stars may be in a phase between T Tauri and post T Tauri stars. More details about that topic can be found in the review of Haisch et al. (1991). So far we have considered only stars which have an activity level by orders of magnitude larger than the Sun.

Other indicators for stellar activity are:

- EUV lines,

- Hα, He $\lambda 10830$Å ,

- H and K lines of Ca II,

- Mg II.

The size and extent of chromospheric active regions varies dramatically over the course of the activity cycle. Thus by measuring the H and K lines of other stars we can infer on stellar activity cycles. One of the programs that is being carried out since a long time is the HK project. Almost 100 stars have been observed continuously since 1966; at present the project is monitoring long-term changes in chromospheric activity for approximately 400 dwarf and giant stars. In order to compare the data with the Sun observations of reflected sunlight from the Moon are done at Mt. Wilson and at Sac Peak and Kitt Peak National Observatory. The sampling of the stars occurs rapidly: usually less than 10 min per star. The accuracy of the instrument is between 1% and 2%. When plotting the HK index against the $B - V$ color index (which is a measure for temperature as explained in chapter 1) then a clear trend can be seen. The HK index increases as the stellar temperature decreases. At this point one must be careful with the interpretation. It is not meant an absolute increase but a relative increase because in cooler stars also the continuum decreases.

In 1972 Skumanich stated the $t^{-1/2}$ law for the time of stellar rotation and stellar chromospheric decay; the rotational velocity and the strength of the CaII emission of a late type star vary inversely with the square root of the star's age. However later it was found that except massive T Tauri stars the majority of low mass stars rotates slowly.

It was also found that there exists a granulation boundary in the HRD at F5 III. Stars of later spectral type begin to develop a convective envelope that grows for the rest of their evolution. At the boundary these envelopes are extremely thin (only 3% of the star's radius). Stars on the right hand side in the HRD of the granulation border have smaller rotation rates.

In hydrodynamics, by definition, the *Rossby Number* is a ratio of inertial forces to the Coriolis Force for a rotating fluid. In astrophysics it is the ratio of the rotation period to the turnover time of the largest convective eddy. In stars with low Rossby numbers the rotation rate dominates the convective turnover time. The low Rossby number correlates well with the strong MgII 1940 emission. A low value of the Rossby number indicates a greater influence of the Coriolis forces. That means that the α effect becomes more important.

Stars can only be observed as point sources since we have no spatial resolution. Some stars show two simultaneous cycle periods. Other stars either have variable activity, or long trends in activity - longer than our 30-year baseline, or appear to be very inactive.

For further details on that topics the book of Schrijver and Zwaan (2000) is recommended where you find further references.

Dravins et al. (1993a) made a detailed comparison of the current sun (G2 V) with the very old solar-type star Beta Hyi (G2 IV) in order to study the post main-sequence evolution of stellar activity and of non thermal processes in solar-type atmospheres. This star has an age of 9.5 +/- 0.8 Gyr. The relatively high lithium abundance may be a signature of the early sub giant stage, when lithium that once diffused to beneath the main-sequence convection zone is dredged up to the surface as the convection zone deepens. Numerical simulations of the 3D photospheric hydrodynamics show typical granules to be significantly larger (a factor of about 5) than solar ones. The emission of the Ca H and K profiles was found to be weaker than that of the Sun. The observations suggest continuous changes in the chromospheric structure, rather than the sudden emergence of growth of active regions (Dravins et al., 1993b)

How can we measure stellar parameters like differential rotation that play a key role in the onset of stellar dynamos? Let us assume we have a rapidly rotating spotted cool star and that it is observed one week apart. By comparing brightness/magnetic images of that star over such time intervals one can measure the rotation rates of starspots at different latitudes over several rotation cycles (Barnes et al 2001).

Since several extrasolar planets have been found one should rise the question whether some of them might be suitable for life. Climatic constraints on planetary habitability were investigated by Kasting (1997). They found such zones around main sequence stars with spectral types in the early F to the mid K-range. The large amount of UV radiation emitted by early type stars poses a problem for evolving life in their vicinity. But there is also a problem with late-type stars; they emit less radiation at wavelengths < 200 nm which is required to split O_2 and initiate ozone formation. The authors show that Earth-like planets orbiting F and K stars may well receive less harmful UV radiation at their surfaces than does the Earth itself.

Chapter 7

The Sun and Climate

In this chapter we discuss the effects of solar variations on the Earth's atmosphere. The variations are small at and above 30 km height but increase to a factor of 2 above 200 km. In general, all correlations of the solar cycle with weather and climate must be treated with great caution, because besides the Sun there are many other influences.

7.1 The Earth's Atmosphere

7.1.1 Structure of the Atmosphere

The Earth's atmosphere is essential to life. It insulates the inhabitants of Earth from the extreme temperatures of space, filters out most radiation dangerous to life etc. It also provides the pressure that is necessary for liquid water at moderate temperatures on the surface. Considering the average temperature profile for the Earth's atmosphere, we can define the following regions:

- Troposphere: characterized by convective motions; warmer air is comparatively light and tends to rise, colder air is dense and tends to sink; the temperature decreases down to 200 K at it's upper boundary, the tropopause, at a height of 17 km. Most of the clouds and weather systems are located in the troposphere.

- Stratosphere: here the temperature slightly increases up to the stratopause at a height of about 50 km. In this layer there are no vertical motions, only horizontal motions occur. If a blub of air tends to rise it immediately becomes colder and thus denser and the buoyancy stops such motions. The temperature in this region increases gradually to -3^0 Celsius, due to the absorption of ultraviolet radiation. The ozone layer, which absorbs and scatters the solar ultraviolet radiation, is located there. Ninety-nine percent of "air" is located in the troposphere and stratosphere. The stratopause separates the stratosphere from the next layer.

Table 7.1: Composition of the Earth's atmosphere

Gas	Molecular weight	fraction by volume [10^{-6}]
N_2	28.02	780900
O_2	32.00	209500
Ar	39.94	9300
CO_2	44.10	300
CO	28.01	0.1
CH_4	16.05	1.52
N_2O	44.02	0.5
H_2O	18.02	$10^4 ... 10^3$

- Mesosphere: the temperature falls down again up to -93^0 Celsius up to the mesopause (height 80 km). The mesopause is the coldest region in the atmosphere.

- Thermosphere (also called ionosphere): the temperature rises up to 1 000 K at a height of 250 km. In this region, thermal conduction is very important. The extension is up to 600 km. The structure of the ionosphere is strongly influenced by the charged particle wind from the Sun (solar wind), which is in turn governed by the level of solar activity. One measure of the structure of the ionosphere is the free electron density, which is an indicator of the degree of ionization.

Tropopause and troposphere are known as the lower atmosphere, stratosphere, stratopause, mesosphere and mesopause are called middle atmosphere and the thermosphere is called the upper atmosphere.

Incoming solar radiation with wavelength larger than 300 nm (in the visible part of the spectrum) penetrates down to the bottom. Radiation with 200 nm < λ < 300 nm is absorbed in the stratosphere (ozone layer) and solar radiation below 100 nm at higher layers.

Generally, the atmosphere extends from the surface down to more than 1 000 km. Up to a height of about 100 km the composition is more or less constant. This is because of the high frequency of collisions between the molecules. These collisions become less efficient at heights above 100 km. The molecules experience a force of gravity that is proportional to their mass. Heavy gases are bound more closely to the Earth and lighter gases flout freely. Hence the lighter atomic oxygen is more abundant at heights above 160 km than the heavier nitrogen N_2.

Sunlight is absorbed in the atmosphere and this process is mainly responsible for its thermal structure. More than 50% of the energy incident from the Sun is absorbed by the surface. 30% is reflected back into space (20% from the clouds, 6% by air and 4% by the surface itself). The atmosphere absorbs only 16% of the incident solar energy. Most of this absorbed energy is captured by dust particles in the troposphere. If we want to construct a model of the atmosphere we have to take into account that it is exposed to two different radiation fields: a) from

7.1. THE EARTH'S ATMOSPHERE

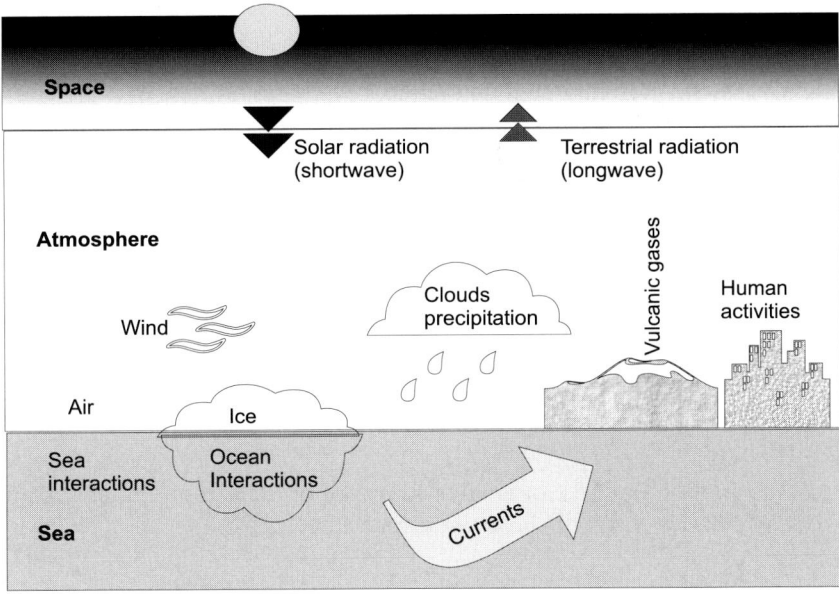

Figure 7.1: Major elements of the climate system

the Sun (covering all wavelengths from far UV to IR), and b) from IR radiation reflected at the surface of the Earth.

The overall heat budget of the atmosphere is as follows: the surface receives 17% of its heat directly from the Sun, 15% from solar radiation scattered by clouds and 68 % from absorption of infrared radiation emitted by the atmosphere. What happens to the energy that is absorbed by the surface? The greater part (79%) is returned to the atmosphere in the form of radiation. The remainder part (21%) is transmitted to the atmosphere by conduction and as by product by the exchange of water H_2O. The surface cools when water evaporates and heat is transmitted to the air as vapor which recondenses to form clouds. Such phase transitions of H_2O play a major role in the energy budget of the lower atmosphere.

The region below 100 km is called *homosphere*, the region above 100 km the *heterosphere*.

7.1.2 Composition

The composition of the Earth's atmosphere is given in Table 7.1.

Of course there are gases that can vary considerably both in space and time like nitric oxide, carbon monoxide and ozone. We can also consider the atmosphere as an extension of the biosphere, especially for gases like O_2, CO_2, CH_4, H_2. Oxygen is produced by photosynthesis:

$$h\nu + CO_2 + H_2O \longrightarrow CH_2O + O_2 \tag{7.1}$$

Table 7.2: Current Greenhouse Gas Concentrations and Other Components

Gas	Pre-industrial conc. (1860)	Present tropospheric conc.	GWP 100 yr time horizon	Atm. lifetime (yr)
CO_2 (ppm)	288	369.4[1]	1	120
CH_4 (ppb)	848[2]	1839[3]/ 1726[4]	23	12
N_2O (ppb)	285[5]	315[3]/ 314[4]	296	114
CCl_3F (ppt)	zero	263[3]/ 260[4]	3,800	50
CF_2Cl_2 (ppt)	zero	544[3]/ 537[4]	8,100	102
$C_2F_3Cl_3$ (ppt)	zero	82[3]/ 82[5]	4,800	85
surface ozone (ppb)	2517	2418/ 2919	20	hours

In this formula CH_2O denotes any variety of organic compounds. Aerobic respiration and decay occur in the reverse reaction:

$$CH_2O + O_2 \longrightarrow CO_2 + H_2O + \text{energy} \quad (7.2)$$

In the absence of this reaction, carbon would accumulate in organic form and the fuel for photosynthesis (atmospheric CO_2) would be depleted. If the supply of O_2 is limited such as in the sediments of organic rich swamps and in the stomachs of ruminants, we get as a product methane CH_4.

In Table 7.2 the change of the greenhouse gas and other gas concentrations of the Earth's atmosphere is given.

The measurements are from:
[1] in situ air samples collected at Mauna Loa Observatory, Hawaii (Keeling, C.D., Whorf, T.P.). [2] : Etheridge, D. M.; Pearman, G. I.; Fraser, P. J. , Tellus, Series B - Chemical and Physical Meteorology, 44B, no. 4, 282. These authors used an ice core from the antarctic called DE08.The extracted ice-core air is analyzed for methane using gas chromatography with flame-ionization detection. The mean air-age was 35 yr younger than the host ice. [3] Values from Macehead, Ireland. [4] Cape Grim, Tasmania [4] data from Law Dome BHD ice core, Etheridge et al.

In Table 7.2 the present tropospheric concentration estimates are calculated as annual arithmetic averages; ppm = parts per million (10^6), ppb = parts per billion (10^9), ppt = parts per trillion (10^{12}).

The Global Warming Potential (GWP) is generally used to contrast different greenhouse gases relative to CO_2. The GWP provides a simple measure of the relative radiative effects of the emissions of various greenhouse gases and is calculated using the formula:

$$GWP = \frac{\int_0^n a_i c_i dt}{\int_0^n a_{CO_2} c_{CO_2} dt} \quad (7.3)$$

where a_i is the instantaneous radiative forcing due to a unit increase in the concentration of trace gas, i, c_i is concentration of the trace gas, i, remaining at time, t, after its release and n is the number of years over which the calculation

Table 7.3: Historical CO_2 record from the Siple Station Ice Core

Depth [m]	Samples measured	Date of ice	Date of Air enclosed	CO_2 conc. (ppmv) in extracted air
187.0-187.3	10	1663	1734-1756	279
177.0-177.3	10	1683	1754-1776	279
162.0-162.3	9	1723	1794-1819	280
147.0-147.2	10	1743	1814-1836	284
128.0-129.0	47	1782	1842-1864	288
111.0-112.0	26	1812	1883-1905	297
102.0-103.0	26	1832	1903-1925	300
92.0-93.0	25	1850	1921-1943	306
82.0-83.0	28	1867	1938-1960	311
76.2-76.6	11	1876	1947-1969	312
72.4-72.7	11	1883	1954-1976	318

is performed. This formula is taken from the Intergovernmental Panel on Climate Change (IPCC) (Houghton et al, 1990).

The CO_2 data are from an ice core analyzed by Neftel et al. (1985). An example of their measurements is given in Table 7.3.

These measurements of the CO_2 gas concentration enclosed in an ice core from Siple Station, Antarctica, indicate that atmospheric CO_2 concentration around 1750 was 280 ± 5 ppmv (parts per million per volume) and has increased since, essentially because of human factors, by 22.5 percent to 345 ppmv in 1984. The natural and anthropogenic changes in atmospheric CO_2 over the last 1000 years from air in Antarctic ice and firn was described in Etheridge et al. (1996).

7.1.3 Paleoclimatology

First of all let us give a definition of climate: Climate is the weather we expect over the period of a month, a season, a decade, or a century. More technically, climate is defined as the weather conditions resulting from the mean state of the atmosphere-ocean-land system, often described in terms of "climate normals" or average weather conditions. Climate Change is a departure from the expected average weather or climate normals.

The reconstruction of past climate changes is one of the great tasks in climate research. Since there exits only a 140 years instrumental record, we have to use proxies to reconstruct climate in the past. Some widely used proxy climate data types are:

- Historical data: Historical documents contain a wealth of information about past climates (diaries, records...)

- Corals: Corals build their hard skeletons from calcium carbonate, a mineral extracted from sea water. The carbonate contains oxygen and the isotopes of oxygen, as well as trace metals, that can be used to determine the temperature of the water in which the coral grew. These temperature recordings can then be used to reconstruct climate during that period of time that the coral lived. Increased sea surface temperature has negative effects on the health of coral. The most visible symptom of declining coral health is coral bleaching.

- Fossil pollen: Each species and genus of plants produces pollen grains which have a distinct shape. These shapes can be used to identify the type of plant from which they came. Pollen grains are well preserved in the sediment layers that form in the bottom of a pond, lake or ocean; an analysis of the pollen grains in each layer tells us what kinds of plants were growing at the time the sediment was deposited. Inferences can then be made about the climate based on the types of plants found in each layer.

- Tree rings: Since tree growth is influenced by climatic conditions, patterns in tree-ring widths, density, and isotopic composition reflect variations in climate. In temperate regions where there is a distinct growing season, trees generally produce one ring a year, and thus record the climatic conditions of each year. Trees can grow to be hundreds to thousands of years old and can contain annually-resolved records of climate for centuries to millennia.

- Ice cores: Located high in mountains and deep in polar ice caps, ice has accumulated from snowfall over many centuries. Scientists drill through the deep ice to collect ice cores. These cores contain dust, air bubbles, or isotopes of oxygen, that can be used to interpret the past climate of that area. Let us briefly discuss one example of isotope measurements: Of the temperature dependent markers the most important is the ratio of ^{18}O to ^{16}O. This can be explained by the fact that water molecules composed of $H_2^{18}O$ evaporate less rapidly and condense more readily than water molecules composed of $H_2^{16}O$. Thus, in the ice cores one obtains annual layers starting with ^{18}O rich, becoming ^{18}O poor, and ending up ^{18}O rich.

  This process also depends on the relative temperatures of different years, which allows comparison with paleoclimatic data.

- Volcanic eruption: After the eruption of volcanoes, the volcanic ash and chemicals are washed out of the atmosphere by precipitation and these eruptions leave a distinct marker within the snow which washed the atmosphere.

  We can then use recorded volcanic eruptions to calibrate the age of the ice-core (here the deuterium to hydrogen ratio is an important proxy).

  Ice cores from Vostok, Antarctica, were the first to cover a full glacial-interglacial cycle.

- Ocean and lake sediments: Between 6 and 11 billion tons of sediment (tiny fossils and chemicals) accumulate in the ocean and lake basins each year.

How can we infer e.g. from ice cores past climate? The accumulation which is governed by saturation water pressure was lower during colder periods and vice

7.1. THE EARTH'S ATMOSPHERE

versa. Accumulation rates inferred in this way are supported by measurements of the cosmogenic isotope Beryllium 10 (^{10}Be), an isotope produced by the interaction of cosmic rays and the upper atmosphere, can be used to determine past snow accumulation in Vostok ice. Deposition of ^{10}Be is assumed to be constant. The chronology of the ice at Vostok has been established down to 2546 m, which is dated at 220 000 years before present. The other two elements which are important are ^{18}O and deuterium. In Antarctica, a cooling of 1°C results in a decrease of 9 per mil deuterium. The last ice age is characterized by three minima separated by slightly warmer episodes called interstadials.

Air initially enclosed in Vostok ice provides our only record of variations in the atmospheric concentrations of CO_2 and CH_4 over a complete glacial-interglacial cycle. For both greenhouse gases, concentrations were higher during interglacial periods than during full glacial periods.

Crowley (2000) discussed the causes of climate change over the past 1000 years. His main conclusion is that as 41-64% of pre-anthropogenic (pre-1850) decadal-scale temperature variations were due to changes in solar irradiance and volcanism.

Several periods of warmth (listed below) have been hypothesized to have occurred in the past. However, upon close examination of these warm periods, it becomes apparent that these periods of warmth are not similar to 20th century warming for two specific reasons:

a) the periods of hypothesized past warming do not appear to be global in extent, or b) the periods of warmth can be explained by known natural climatic forcing conditions that are uniquely different than those of the last 100 years.

Examples of periods of warmth:

- Medieval: ~ 9th to 14th centuries; this seems to be in doubt now because the temperature anomaly at that time was very small; however the Little Ice Age for the northern hemisphere from 15th to 19th centuries is clearly seen (Fig. 7.2).

- mid-Holocene warm Period (approx. 6 000 years ago); this seems to be in connection with changes of the Earth's orbit (Theory of Milankovich).

- Penultimate interglacial period (approx. 125 000 years ago). It appears that temperatures (at least summer temperatures) were slightly warmer than today (by about 1 to 2°C), for reasons that are well known - the changes in the Earth's orbit (Hughes and Diaz, 1994).

- Mid-Cretaceous Period (era?) (approx. 120-90 million years ago): Breadfruit trees apparently grew as far north as Greenland (55 N), and in the oceans, warm water corals grew farther away from the equator in both hemispheres. The mid-Cretaceous was characterized by geography and an ocean circulation that was vastly different from today, as well as higher carbon dioxide levels (at least 2 to 4 times higher than today).

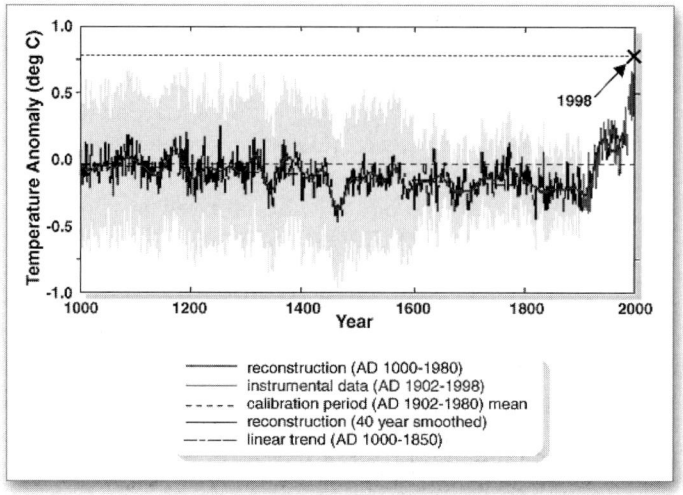

Figure 7.2: Temperature anomaly clearly showing the Little Ice Age

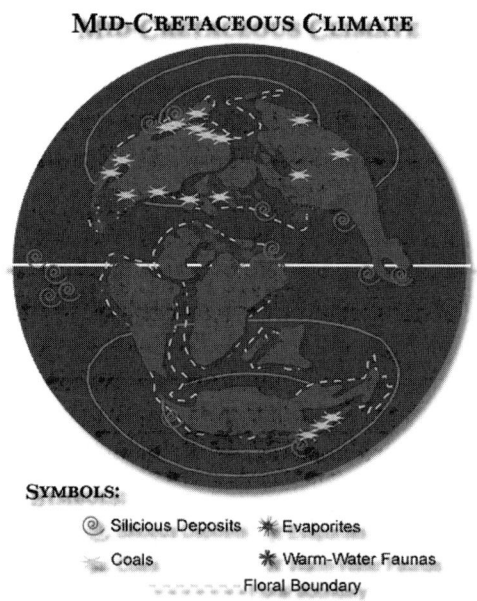

Figure 7.3: Cretaceous climate and land/sea distribution; Image credit: Crowley and North, 1991

7.1. THE EARTH'S ATMOSPHERE

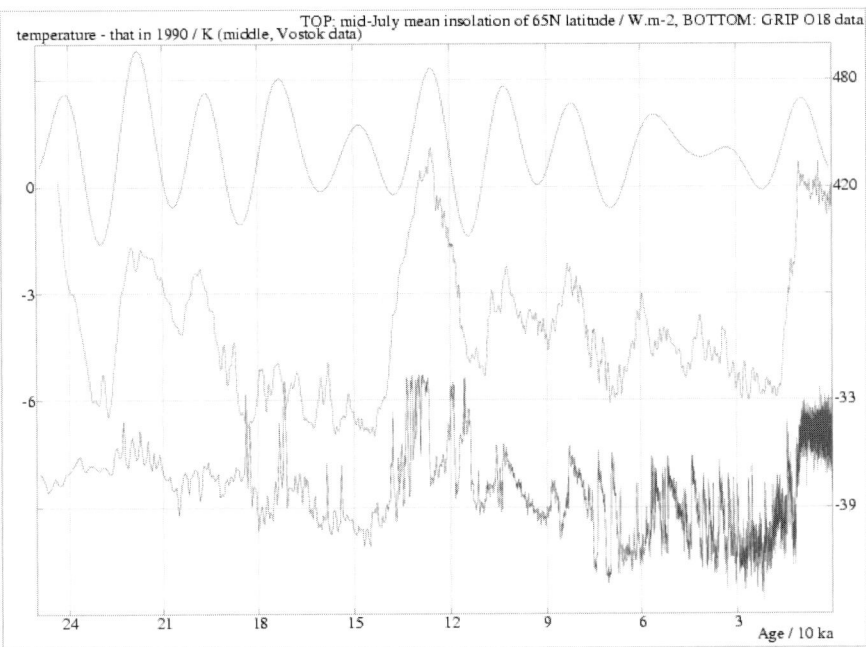

Figure 7.4: Upper curve: average insolation of 65 degrees northern latitude (Watts per one square meter of a horizontal atmosphere) in mid-July. As seen, it varies from some 390 to 490 W/m². Middle curve: Global temperature (Vostok ice core). Lower Curve: Greenland, GRIP core. Image courtesy: Jan Hollan

7.1.4 Theory of Milankovich

Seasons on Earth are caused by the tilt of the Earth's rotation axis relative to its plane of revolution around the Sun (which is called the ecliptic). In summer, one hemisphere is pointing toward the Sun, at the same time the opposite hemisphere is in winter. If the Earth's axis were not inclined every point on the earth would receive the same amount of sunlight each day of the year. Changes in this tilt can change the severity of the seasons. More tilt means more severe seasons, i.e. warmer summers and colder winters. The tilt of the Earth's axis changes between 22 and 25 degrees on a cycle of about 41 000 years. If the summers are cool than snow and ice last from year to year in high latitudes building up massive ice sheets. Now positive feedbacks in the climate system start to work. Snow reflects more of the sun's energy into space causing additional cooling. Also the amount of CO_2 falls as ice sheets grow and thus adding to the cooling.

Another astronomical effect on climate is that the orbit of the earth is not circular. Presently, perihelion (closest approach to the Sun) occurs in January, thus on the northern hemisphere winters are slightly milder. The perihelion changes in a cycle of 22 000 years. Therefore, 11 000 years ago perihelion occurred in July making seasons more severe than today. The eccentricity of the earth's orbit varies

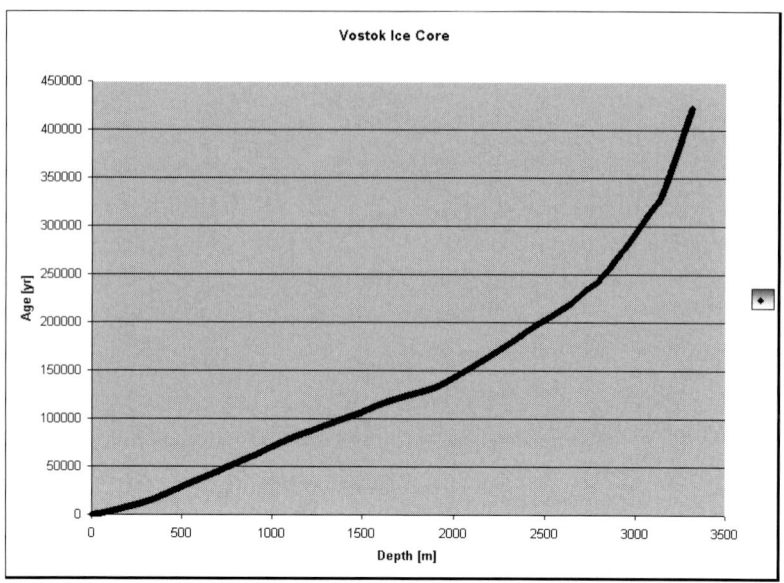

Figure 7.5: Vostok Ice core. Different depth can be attributed to different ages

on cycles of 100 000 and 400 000 years. It is the combined effect of the 41 000 year tilt cycle and the 22 000 year perihelion cycle plus the small effect from the eccentricity that influences the climate. These variations of the Earth's orbit were first investigated by Milankovich.

To study the effect of these astronomical variations on climate one must take into account, that orbital changes occur over thousands of years and the climate system also takes thousand of years to respond. The primary driver of ice ages seems to be the total summer radiation received in northern latitude zones near 65^0 north (this is where the major ice sheets formed in the past) and past ice ages correlate with the 65N summer insolation. Astronomical calculations show that the 65N summer insulation should increase gradually over the next 25 000 years. No decline of the 65N summer insolation that is sufficient to cause an ice age is expected within the next 100 000 years.

The most important sources of information about such changes and the associated composition of the atmosphere are the two large ice caps of Greenland and Antarctica. Analysis of ice cores is the most powerful means we have to determine how climate has changed over the last few climatic cycles. The concentrations of the principal greenhouse gases CO_2, CH_4, N_2O played an important role in the transitions from cold ice age climates to warmer interstadials. Warm interstadials have always been accompanied by an increase of the atmospheric concentration of the three principal greenhouse gases. This increase has been, at least for CO_2, vital for the ending of glacial epochs. A highly simplified course of events for the past four transitions would then be as follows:

- changing orbital parameters initiated the end of the glacial epoch

7.1. THE EARTH'S ATMOSPHERE

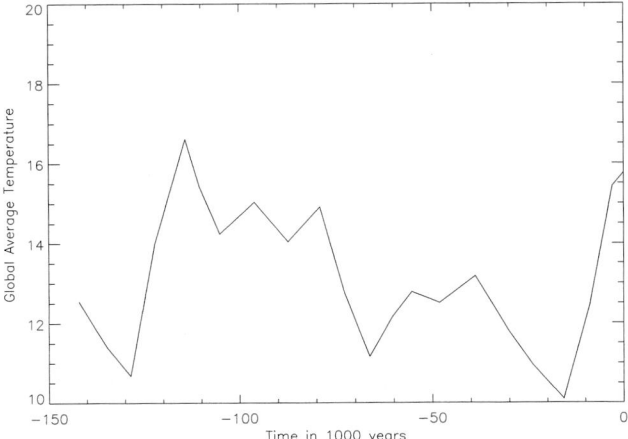

Figure 7.6: Variation of global temperature over the last 150 000 years

- an increase in greenhouse gases then amplified the weak orbital signal

- in the second half of the transition, warming was further amplified by decreasing albedo, caused by melting of the large ice sheets in the Northern Hemisphere going parallel with a change of the ocean circulation.

The isotopic records of Greenland ice cores show evidence for fast and drastic climatic changes during the last glacial epoch. Possible causes and mechanisms of such changes and their significance as global climatic events are discussed by Stauffer (2000). Ice core results also enable the reaction of the environment to past global changes to be investigated. The deglaciation of the northern hemisphere is described in Alley and Clark (1999). A carbon cycle model was used to reconstruct the global mean surface temperature during the last 150 Million years showing that during this period the tectonic forcing such as decrease in volcanic activity and the formation and uplift of the Himalayas and the Tibetan Plateau dominated the control of the climate (Tajika, 2001).

7.1.5 Greenhouseffect

Trace constituents of the atmosphere such as H_2O, CO_2, O_3 absorb energy at longer wavelengths and thus trap heat radiated by the surface. The effect is very similar to that of a glass pane in a greenhouse. The atmosphere is transparent to solar radiation but it is opaque to longer wavelengths. The infrared absorbing gases return heat to the ground and account for about 70% of the net input of energy to the surface. If our atmosphere would contain no water vapor and carbondioxide, the surface temperature would be about 40 K colder than today. This would imply that large portions of the planet would be covered with ice.

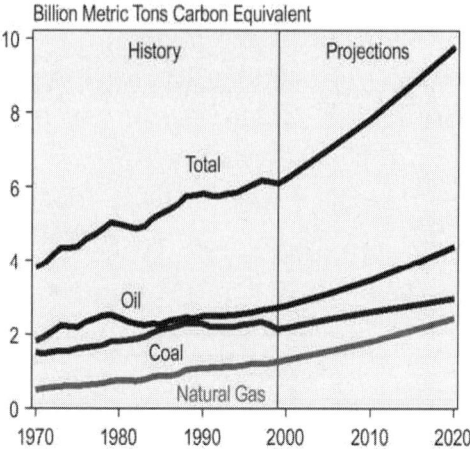

Figure 7.7: World Carbon Dioxide Emissions (US Dept. of Energy)

Since the 1980s there is a growing concern that the increase in the abundance of carbondioxide caused by combustion of fossil fuels could lead to a general warming of the global climate (see Fig. 7.7). Similar greenhouse effects arise from the gases methane, nitrous oxide and chlorofluorcarbons (CFCs). All these gases are referred to as greenhouse gases due to their ability to trap heat.

7.1.6 Ozone

The absorption of shortwave solar radiation in altitudes above the troposphere is responsible for the temperature increase in these layers. Ozone absorbs most of the UV portion of sunlight ($200 < \lambda < 300$ nm). The absorption process results in the dissociation of O_3. The recombination process involves the reactions:

$$O + O_2 \rightleftharpoons O_3^* \tag{7.4}$$
$$O_3^* + M \rightarrow O_3 + M \tag{7.5}$$

O_3^* denotes an unstable higher energetic level of O_3. By collision with other atmospheric molecules M this energy excess is removed. The rate of recombination, i.e. the number of O_3 molecules formed per unit volume per unit time is proportional to the product of:

- O
- O_2
- M

7.1. THE EARTH'S ATMOSPHERE

We can make the simplification that throughout the stratosphere all of the radiative energy from the sun that is absorbed by O_3 is converted locally to heat. The heating rate depends on the distribution of Ozone with height and on the incoming solar energy.

Tropospheric ozone is either produced by oxidation of hydrocarbons and CO or by downward transportation of stratospheric ozone. Some examples of reactions are given below.

$$\begin{aligned} CO + OH &\rightarrow CO_2 + H \\ H + O_2 + M &\rightarrow HO_2 + M \\ HO_2 + M &\rightarrow OH + NO_2 \\ h\nu + NO_2 &\rightarrow NO + O \\ O + O_2 + M &\rightarrow O_3 + M \\ CO + 2O_2 &\rightarrow CO_2 + O_3 \end{aligned}$$

Natural events such as Volcanic Eruptions can strongly influence the amount of Ozone in the atmosphere. However, man-made chemicals such as CFCs or chlorofluorocarbons are now known to have a very dramatic influence on Ozone levels too. CFCs were once widely used in aerosol propellants, refrigerants, foams, and industrial processes. Changes in the ozone layer caused by release of CFC's in the atmosphere have the potential of producing biological damage through increased UVB radiation. While cloud cover provides protection on the ground against solar radiation in the visible and near UV wavelengths, biologically damaging radiation near 300 nm is controlled primarily by the total ozone content.

The ozone is measured in Dobson units. 1 Dobson Unit (DU) is defined to be 0.01 mm thickness at STP (standard temperature and pressure). Ozone layer thickness is expressed in terms of Dobson units, which measure what its physical thickness would be if compressed in the Earth's atmosphere. In those terms, it's very thin indeed. A normal range is 300 to 500 Dobson units.

In the Earths lower atmosphere, near ground level, ozone is formed when pollutants emitted by cars, power plants, industrial boilers, refineries, chemical plants, and other sources react chemically in the presence of sunlight. Ozone at ground level is a harmful pollutant. Ozone pollution is a concern during the summer months, when the weather conditions needed to form it, lots of sun, hot temperatures, normally occur.

7.1.7 The Structure of the Higher Atmosphere

Temperature inversion in the thermosphere

Above 80 km there is an inversion of the temperature that is caused by the absorption of solar radiation below 200 nm. Let us briefly discuss the most important processes:

- $100\,nm < \lambda < 200\,nm$: absorption of solar radiation leads to a dissociation of O_2:
$$h\nu + O_2 \to O + O \qquad (7.6)$$

- shorter wavelengths: ionization of O, O_2, N_2:
$$\begin{aligned} h\nu + O &\to O^+ + e \\ h\nu + O_2 &\to O_2^+ + e \\ h\nu + N_2 &\to N_2^+ + e \end{aligned}$$

The electrons that are emitted by these reactions loose energy by collision, elastic and inelastic. This can cause further ionization and contribute to the production of excited states and the associated emission of airglow.

Electrons can be removed by dissociative recombination:
$$\begin{aligned} e + O_2^+ &\to O + O \\ e + NO^+ &\to N + O \end{aligned}$$

There is of course a balance between dissociation of O_2 and recombination.

- Reformation of molecular oxygen:
$$\begin{aligned} O + O_2 &\to O_3 + M \\ H + O_3 &\to OH + O_2 \\ O + OH &\to O_2 + H \end{aligned}$$

In the first of these reactions the recombination of oxygen is catalyzed by the presence of hydrogen. Such catalytic reactions play an important role in the chemistry of the atmosphere below 80 km. The density in the thermosphere is low, therefore O diffuses downward. The recombination requires higher densities and is confined to regions below 100 km. The dissociation of O_2 can occur at any level.

Hydrogen loss

Any particle in the atmosphere is bound to the Earth by the force of gravity. If we move such a particle a vertical distance Δz then the work $mg\Delta z$ is done. m denotes the mass of the particle, g the gravitational acceleration $=9.81\,\mathrm{m/s^2}$. The work that must be done to escape the gravitational field is mgR, where R is the radius of the Earth $\sim 6\,400$ km. All atoms or molecules have a range of speeds that is described by the Maxwell–Boltzmann distribution. The average kinetic energy is given by:
$$E_{kin} = \frac{3}{2}kT \qquad (7.7)$$

where k is the Boltzmann constant ($1.38 \times 10^{-16}\,\mathrm{erg\,K^{-1}}$). Thus an atom can escape the gravitational field if its thermal kinetic energy $\sim kT$ is much larger than mgR. Of course we must also consider collisions (except at the highest level in the atmosphere). At the high temperatures in the thermosphere (700...2 000 K), significant numbers of hydrogen atoms have velocities above the escape velocity $v_{esc} \sim 11.2$ km/s. Therefore, hydrogen is lost at a rate of 10^8 atoms per cm^2 per second. These escaping hydrogen atoms are derived mainly from the oceans and over the past 4.5×10^9 years of the Earth's history, the sea level has declined by two meters globally. Of course during this reaction also O_2 is set into the atmosphere which was crucial for the evolution of life.

There is also a significant of loss of helium.

7.2 Earth's History and Origin of the Atmosphere

In this section we discuss the main features in the evolution of the Earth and life on the Earth.

7.2.1 History of the Earth

The history of the Earth can be mainly divided into four parts:

- Precambrian: Earth's history up to 570 million years ago.
- Paleozoic: 570-240 million years ago,
- Mesozoic: 240-65 million years ago,
- Cenozoic: 65 million years ago to present.

Precambrian Time

Between 4 and 2.5 billion years ago the continents, atmosphere and oceans formed, as well as one celled organisms known as prokaryotes. These are the ancestors of present day bacteria and cyanobacteria. The atmospheric oxygen increases and later multicellular organisms appear.

Paleozoic Era

This era is subdivided into:

- Cambrian Period: 570-500 million years ago; multicellular life.
- Ordovician Period: 500-435 million years ago; primitive life on land; vertebrates in the ocean.
- Silurian Period: 435-410 million years ago; first plants and insects appear on land.
- Devonian Period: 410-360 million years ago; spiders, mites and amphibians.
- Carboniferous Period: 360-290 million years ago; first true reptiles appear; coal begins to form.

- Permian Period: 290-250 million years ago; mysterious mass extinction of life; 90 % of all organisms die out; reptiles inherit the Earth.

Mesozoic Era

This era is divided into:

- Triassic Period: 240-205 million years ago; small dinosaurs, ichthyosaurs, plesiosaurs, first true mammals

- Jurassic Period: 205-138 million years ago; huge dinosaurs, flying pterosaurs, oldest known birds.

- Cretaceous Period: 138 -65 million years ago: global warming, spread of dinosaurs. Sudden mass extinction (probably due to asteroid impact), 70 % of all organisms die out at the end of this period.

Cenozoic Era

This is divided into a Tertiary and a Quaternary Period. The Tertiary Period is further divided into:

- Paleocene Epoch: 65-55 million years ago; mammals inherit Earth.

- Eocene Period: 55-38 million years ago; ancestral forms of the horse, rhinoceros, camel and other modern groups such as bats, primates etc. Mammals adapt to marine life.

- Oligocene Epoch: 38-24 million years ago; elephants, cats and dogs, monkeys.

- Miocene Epoch: 24-5 million years ago; global climate cools; establishment of the Antarctic ice sheet; large apes in Africa and southern Europe.

- Pliocene Epoch: 5 -1.6 million years ago; climate becomes cooler and drier. Mammals are well established as the dominant terrestrial life form; ancestors of modern humans.

The Quaternaray Period is divided into:

- Pleistocene Epoch: 1.6 million to 10 000 years ago; most recent global ice age; glacier ice spreads out over more than one-fourth of Earth's land surface; modern humans arise and begin their migrations.

- Holocene Epoch 10 000 years ago to present; global climate moderates; ice sheets retreat from Europe and North America; sea levels rise.

7.2.2 Origin of the Atmosphere

Let us start with the remark that the origin of our earth's atmosphere is still subject to much speculation. However the most probable history of its evolution was as follows.

Our Earth was formed some 4.5 billion years ago. At that time it was probably too hot to retain any primordial atmosphere. This first atmosphere most probably consisted of helium, hydrogen, ammonia and methane. At that time the Earth was a very active planet from the geologic point of view. Volcanism was widespread and if we assume that volcanoes five billion years ago emitted the same gasses as they do today, the earth's second atmosphere probably consisted of water vapor, carbon dioxide, and nitrogen. These gasses were expelled from the earth's interior by a process known as outgassing.

It is also possible that the impact of comets brought significant amounts of water and other volatile gasses to the Earth. The vast amounts of water vapor expelled by the volcanic earth resulted in the formation of clouds which, in turn, produced rain. Over a period of thousands of years, the rain accumulated as rivers and lake and ocean basins. This process was extremely important for the carbon dioxide CO_2. The water reservoirs acted as sinks for that gas and through chemical and later biological processes it became locked up in sedimentary rocks as limestone.

On the other hand nitrogen, which is not very chemically active, continued to accumulate in the atmosphere.

What about the most important gas oxygen we need to live? The first oxidized rocks found in geological strata date back only 1.2 billion years (which is quite recently compared to the 4.6 billion year age of the Earth). 600 million years ago oxygen constituted only 1% of the atmosphere (currently 21%). However, oxygen was only a trace gas in the air when life first appeared on the planet. That was one of the reasons that life first evolved in the oceans. Single-celled bacterium dwelling in the oceans did not need oxygen to live. Oxygen first appeared in the environment when early bacteria developed the ability to split water molecules apart using the energy of sunlight - a key part of photosynthesis. Photosynthesizing organisms produced the oxygen that accumulated over geologic time.

These processes acting sequentially and simultaneously appear to have produced the delicate balance of 78% nitrogen (N_2) and 21% oxygen (O_2) we observe today. By the way, oxygen is the third most abundant element in the universe and makes up nearly half of the mass of the Earth's crust, two thirds of the mass of the human body and nine tenths of the mass of water.

The Earth cannot sustain more than $\sim 20\%$ O_2 in the atmosphere. Otherwise spontaneous fires would occur that would deplete the oxygen.

The enrichment of oxygen in the atmosphere might be seen in context with the methane content. Microbes who utilize photosynthesis existed on Earth half a billion years or more before oxygen became prevalent, without substantially affecting the composition of the atmosphere. The transition to an atmosphere with noticeable oxygen content occurred about 2.4 billion years ago. What really happened 2.4 billion years ago which transformed the Earth's atmosphere? According to

Catling et al (2001) after photosynthesis separated the oxygen from the hydrogen, the authors argue, the two components followed separate paths. The free oxygen remained in the Earth's crust, while the hydrogen went on to combine with carbon in a process known as "methanogenesis," producing methane. When methane travelled to the upper atmosphere, ultraviolet radiation from the Sun dissolved it into its components. The light hydrogen drifted away into space and was lost forever to the Earth's atmosphere.

Because the hydrogen was lost while the oxygen stayed on Earth, an excess of oxygen gradually accumulated. When the Earth's crust was saturated, the oxygen spilled out and flooded the ancient atmosphere, creating the oxygen rich environment we know today. This can also solve the faint young Sun problem (see next chapter).

Of course these facts are extremely important to answer questions like:

- How did life begin and first evolve on Earth?

- How have conditions on Earth over the past four billion years changed and affected life?

- What are the most extreme conditions under which life can be found on Earth?

- Where else in our own and other planetary systems have conditions possibly been suitable for life?

- How should one search for evidence of fossil or living microbes at extraterrestrial sites such as Mars or Jupiter's moon Europa?

The field of Astrobiology tries to find answers to these questions.

7.3 The Faint Young Sun

7.3.1 Introduction

According to theories of stellar evolution, the solar constant is not a constant but has been increasing continuously throughout the main sequence lifetime of the Sun. The increase in luminosity can be explained by the conversion of hydrogen into helium; by this energy generation the mean atomic weight and density of the Sun is increased. This produces higher core temperatures and therefore the rate of fusion reactions increases and hence the luminosity. A very rough formula for the luminosity change of the Sun during its main sequence evolution was given by Gough (1981):

$$L(t) = [1 + 0.4(1 - t/t_0)]^{-1} L_o \qquad (7.8)$$

In this formula L_0 is the present solar luminosity and t_0 the present age of the Sun (4.6 Gyr). Other explanations of a possible different solar luminosity at the early evolution of the Sun are:

- Revisions in the standard solar model in order to solve the neutrino problem.

7.3. THE FAINT YOUNG SUN

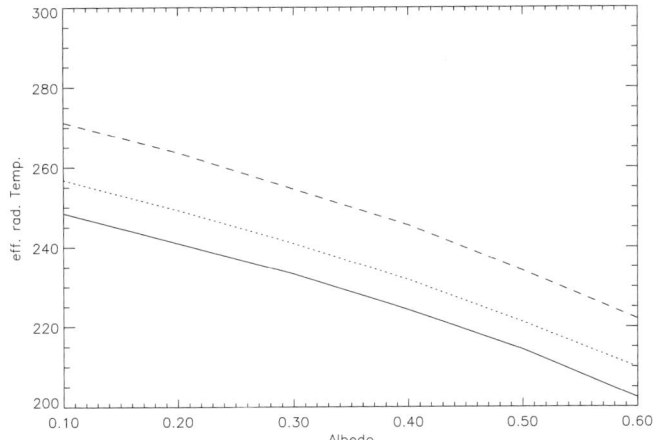

Figure 7.8: Effective radiating temperature of the Earth as a function of planetary Albedo A for three different values of the solar constant, a) 982, b) 1088 (dotted), c) present value 1360 (dashed).

- Strong mass loss during the early phase (Willson et al. 1987).

For our investigations it is clear that a change in solar luminosity over time would have affected the Earth's radiation balance and, thus, its climate. If the Earth is considered to radiate like a blackbody, S is the solar constant (at present 1360 W/m^2), σ the Stefan-Boltzmann constant, A the planetary Albedo (~ 0.3), T_e the effective radiating temperature can be obtained by:

$$T_e = [S(1-A)/4\sigma]^{1/4} \qquad (7.9)$$

The relevant albedo to use here is the *Bond Albedo*, which is the percentage of the total incident solar radiation/reflected back into space. The present effective radiating temperature of the Earth is ~ 255 K. If we combine 7.8 and 7.9 then the increase of T_e was about 20 deg over geologic time if the albedo of the Earth is assumed to remain constant. We must also take into account the Earth's mean surface temperature T_s and

$$T_s > T_e \qquad (7.10)$$

Because of the greenhouse effect the difference between T_s and T_e is about 33 K. The greenhouse effect is caused by the difference in opacity in the visible and infrared regions of the electromagnetic spectrum. The Earth's atmosphere is relatively transparent to incoming solar radiation, but absorbs a large fraction of the outgoing IR. Most of the absorption is caused by the vibration-rotation bands of H_2O and CO_2 and to the pure rotation band of H_2O.

Table 7.4: Typical values for the albedo.

	Albedo
Tropical forest	0.13
Woodland	0.14
Grassland	0.20
Stony desert	0.24
Sandy desert	0.37
Sea ice	0.25-0.60
Snowy ice	0.80
Cool water	<0.08
Warm water	< 0.10

In Figure 7.8 the effective radiating temperature of the earth T_e as a function of planetary albedo is given for three different values of the solar constant, a) at present b) reduced by 20 % and c) reduced by 30 %.

We clearly see that T_e strongly depends on the solar constant and on the Albedo. A larger value of the albedo leads to a lower value of the effective radiating temperature, the Earth becomes cooler. The albedo can increase because of:

- increased glaciation of the Earth,

- increased fraction of clouds.

Some typical values for A are given in Table 7.4. Sagan and Mullen (1972) first pointed out the implications of this change of solar luminosity for the Earth's climate. Using a very simple model of the greenhouse effect they showed that lower solar luminosity would have resulted in T_s below the freezing point of water for roughly the first 2 Gyr of the Earth's evolution. However this cannot be correct. Already Sagan and Mullen pointed out the presence of pillow lavas, mud cracks and ripple marks in 3.2 Gyr old rocks suggesting strongly the presence of liquid water on the Earth's surface at that time. We also know that sedimentary rocks have been deposited about 3.8 Gyrs ago and these must have formed in liquid water.

This discrepancy between the prediction of simple climate models and the actual climate record has been known as the faint Sun paradox.

One possible explanation might be that the Earth's albedo was significantly lower in the past or that the greenhouse effect of its atmosphere was larger. However, as Sagan and Mullen pointed out a large change in the Earth's albedo was unlikely; any decrease in cloudiness that might result from lower surface temperatures would likely be compensated by an increase in snow and ice cover. However if the Earth's surface was mostly water covered this argument does not work.

From climate research we know that there was no glaciation on Earth prior to about 2.7 Gyr ago (e.g. oxygen isotopes imply warm surface temperatures throughout the Precambrian (Kasting, 1989)).

7.3.2 The CO_2 Geochemical Cycle

Let us start with some numbers: the total surface reservoir of carbon is about 10^{23} g. This is enough to produce a CO_2 partial pressure of about 60 bar were all of it present as gaseous (Holland, 1978). Most of the carbon is contained in carbonate rocks on the continents. A much smaller amount is present in the oceans as carbonate and bicarbonate ions (4×10^{19} g). Presently about 7×10^{17} g are present in the atmosphere (this number is growing). There is an equilibrium between the ocean and the atmosphere at timescales of about 1000 y:

CO_2 is removed from the atmosphere/ocean by weathering of silicate rocks on the continents and 20 % of atmospheric CO_2 is removed by photosynthesis followed by burial of organic carbon. If one represents silicate rocks by $CaSiO_3$ (wollastonite) then the CO_2 loss process can be described by the following three reactions:

$$CaSiO_3 + 2CO_2 + H_2O \rightarrow Ca^{++} + 2HCO_3 + SiO_2 \qquad (7.11)$$
$$Ca^{++} + 2HCO_3 \rightarrow CaCO_3 + CO_2 + H_2O \qquad (7.12)$$
$$(7.13)$$

Thus:

$$CaSiO_3 + CO_2 \rightarrow CaCO_3 + SiO_2 \qquad (7.14)$$

When old sea floor is subducted and carbonate sediments are subjected to higher temperatures and pressures CO_2 is returned to the atmosphere/ocean. Then reaction 7.14 goes in the opposite direction, calcium silicate is reformed and gaseous CO_2 is released. Much of this CO_2 escapes through volcanoes. That process is termed carbonate metamorphism and on the young Earth the rate of carbonate metamorphism could have been augmented by faster rates of tectonic cycling and by impact processing of carbonate rich sediments.

It is important to note that the rates of the weathering reactions are strongly dependent on temperature. The reaction rates increase with temperature and weathering requires liquid water. The temperature dependence of the silicate weathering process rate leads to a negative feedback between atmospheric CO_2 and surface temperature: if the surface temperature were to decrease (because of a faint young Sun), the weathering rate would also decrease and carbon dioxide would begin to accumulate in the atmosphere. This increase of CO_2 causes an increase in the greenhouse effect and thus the temperature increases. The reverse would happen if the climate became warmer: the weathering rate would increase, pCO_2 would decrease and the greenhouse effect would become smaller (Walker et al. 1981). This mechanism can explain why the temperature on Earth was high enough for liquid water even when the solar luminosity was smaller.

The modern rate of CO_2 release from volcanoes would create a 1-bar CO_2 atmosphere in only 20 Myr if carbonates were not forming. This shows that the response time of the system is quite fast in geologic terms.

7.3.3 Effects of the Biota

Presently the CO_2 geochemical cycle is modulated by the biota. Calcium carbonate formation can be largely attributed to the secretion of shells by plankton and other marine organisms. Land plants enhance silicate weathering rates by pumping up the carbon dioxide partial pressure in soils by a factor of 10 to 40 over the atmospheric value; photosynthesis on land and in the oceans creates organic carbon which is then buried in sediments. Thus the atmospheric CO_2 level is reduced. Therefore, the Earth today is probably cooler than it would be in the absence of life. Lovelock (1979, 1988) created therefore the Gaia Hypothesis, which means that the Earth's climate is controlled by biota and would have become unstable where it not for the homeostatic modulation of climate by organisms. Let us assume biological control of the Earth's climate in more detail. According to Berner et al (henceforth BLAG), the dependence of the silicate weathering rate f_w on surface temperature T can be written as:

$$f_w = 1 + 0.087(T - T_0) + 0.0019(T - T_0)^2 \qquad (7.15)$$

Here, T_0 is the present mean surface temperature (288 K) and $f_w = 1$. The CO_2 greenhouse effect is parameterized in the BLAG model as:

$$T - T_0 = 2.88 \ln(P/P_0) \qquad (7.16)$$

where P indicates atmospheric pCO_2 and P_0 the present CO_2 partial pressure. From laboratory studies we know that the weathering rate of silicate minerals varies approximately as

$$pCO_2^{0.3} \qquad (7.17)$$

for CO_2 partial pressures of 2 to 20 bar and temperatures of 100 to 200° C. These data were derived by Lagache (1976) and Walker et al. (1981). Let us assume that we can apply this relation to the Earth's surface conditions. To study the maximum effect let us further assume that removing land plants from the system would reduce surface soil pCO_2 by a factor of 40. Then the equation for the silicate weathering process can be written as:

$$\begin{aligned} f_w &= [1 + 0.087(T - T_0) + 0.0019(T - T_0)^2] \\ &= [P_S/40P_0)]^{0.3} \end{aligned} \qquad (7.18)$$

Here P_S is the partial CO_2 pressure in the soil and today we have $P_S = 40P_0$ and obtain $f_w=1$. On a vegetation free Earth $P_S \sim P_0$, and f_w would be reduced by a factor of $40^{-0.3} \sim 1/3$. The carbon cycle is only balanced when $f_w = 1$. Therefore, without vegetation, the atmospheric pCO_2 and surface temperature would have to increase to bring back the silicate weathering rate to its present value. We substitute equation 7.16 into equation 7.18 and solve for P/P_0 and obtain:

$$P/P_0 = 9 \qquad T - T_0 \sim 6.3 \, \text{K} \qquad (7.19)$$

This shows that under the assumption that land plants pump up soil CO_2 by a factor of 40, the effect of eliminating them would be to increase the Earth's

Table 7.5: Effects of Solar Radiation at different wavelengths on the Middle and Upper Atmosphere

Wavelength [nm]	Variab. middle Atm.	Variab. upper Atm.	Effect	Height [km]
1-10, SXR		sporadic	Ion. all	70-100
10-100, XUV	2ppm	2 x	Ion. N_2, O, O_2	
100-120, EUV	6 ppm	30%	Ion. NO	80-100
120-200, VUV	150 ppm	10%	Diss. O_2	40-130
200-240, UV	0.12%	5%	Diss. O_2, O_3	20-40
240-300, UV	1.0%	<1%	Diss. O_3	20-40

temperature by only 6 deg. The net cooling effect of the biota should be somewhat larger because of the influence of the organic carbon cycle; today 20% of the carbon is organic carbon rather than carbonate. One can estimate that if life suddenly were eliminated in total the temperature would increase by 8 deg. Thus even a lifeless Earth would apparently be no warmer than the real Earth was during the Cretaceous, when the dinosaurs flourished.

The studies of Schwartzmann and Volk (1989) showed that biota may accelerate chemical weathering by stabilizing soil (silicate minerals stay in contact with carbonated water), generating organic acids. This could lead to enhanced weathering rates of up to 1000 instead of 3. Therefore, the CO_2 partial pressure on a lifeless Earth might be as high as a few tenths of a bar and the surface temperature may be up to 60 K warmer!

7.4 The Atmosphere's Response to Solar Irradiation

7.4.1 Introduction

The principal effects of solar radiation on the middle and upper atmosphere are summarized in table 7.5. From that table it follows that the size of variation depends on the wavelength of the solar radiation: it becomes smaller at longer wavelengths. Above 300 nm it is very difficult to detect and can be measured only with satellite radiometric detectors. In addition to radiation the Sun also emits the solar wind which consists of particles that interact with the geomagnetic field to form the Earth's magnetosphere. We observe a large input of electrons and protons (causing the aurora) and ionospheric currents are produced causing joule heating. In principle these phenomena are concentrated at high geomagnetic latitudes; heating effects can spread equatorward by convection and conduction.

The typical structure of the Earth's atmosphere was already shortly described. The boundaries of the various layers (Troposphere, Stratosphere, Mesosphere, Ionosphere) are called pauses (e.g. the Tropopause) and are defined by minima or maxima of the temperature profile. At 100 km the density is 10^{-6} of its surface value. The temperature in the thermosphere is strongly dependent on solar activity. The major sources for heating at this layer are:

- solar ionizing photons,
- magnetospheric processes.

The principal part of the ionosphere is produced by XUV which is strongly absorbed there. Ionization and recombination occurs and this contributes to the heating of the thermosphere. By comparing with table 7.5 we see that the energy involved is small; dissociation of O_2 is strong, above 120 km oxygen occurs as atoms. Through vertical mixing the ratio $O_2 : N_2$ is constant near 0.1 throughout the lower thermosphere. Oxygen atoms are produced down to 30 km and most of them combine with O_2 to form ozone. This attains a peak ratio of 10^{-5} near 30 km. Through this recombination the middle atmosphere is heated (peak at 50 km). Photons around 300 nm can reach the surface. They produce electronically excited oxygen which drives a large fraction of urban pollution chemistry.

7.4.2 Solar Variability

Solar variability can be divided into three components according to their influence on the structure and composition of the atmosphere:

- variation of the solar constant
- XUV and UV variation
- energetic particle variation

Let us consider these components in detail.

Radiative component

So far we have only the discussed the long term solar variability- summarized as the faint young Sun problem and the influence of the changing parameters of the Earth's orbit on climate (Berger, 1980).

We now address to the question whether there exists also a variability of the solar input on shorter timescales. The total solar irradiance describes the radiant energy emitted by the sun over all wavelengths that falls vertically each second on 1 square meter outside the earth's atmosphere. This is the definition of the solar constant. Because of the influences of the Earth's atmosphere this constant is extremely difficult to measure on the surface and the most reliable measurements can only be done from space. In Table 7.6 the satellite measurements and the respective time spans of the measurements are summarized.

The VIRGO Experiment on the ESA/NASA SOHO Mission has two types of radiometers to measure total solar irradiance (TSI): DIARAD and PMO6V. A description of the instrument can be found in Fröhlich et al. (1995). Let us shortly describe the DIARAD measurement facility which is a part of SOHO/VIRGO: DIARAD is a Differential Absolute Radiometer. It is composed of two cylindrical

7.4. THE ATMOSPHERE'S RESPONSE TO SOLAR IRRADIATION

Table 7.6: Satellite measurements of the solar constant

NIMBUS-7	16 Nov 78-13 Dec 93
SMM (ACRIM I)	16 Feb 80-01 Jun 89
ERBS	25 Oct 84-
NOAA-9	23 Jan 85-20 Dec 89
NOAA-10	22 Oct 86-01 Apr 87
UARS (ACRIM II)	5 Oct 91-
SOHO/VIRGO	18 Jan 96-

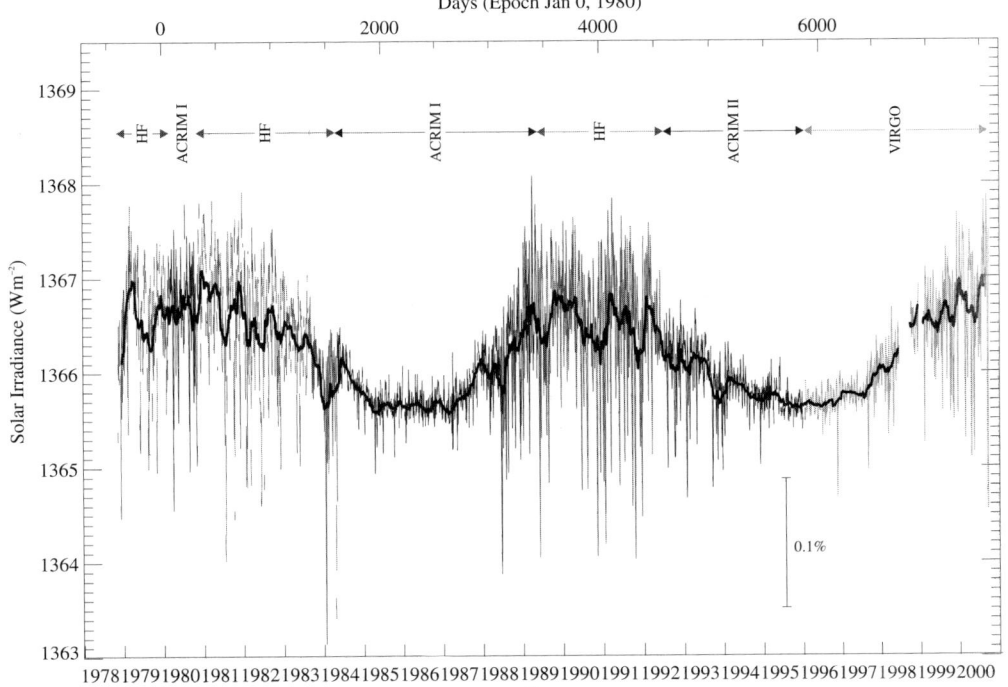

Figure 7.9: Solar irradiance measurements from satellites

cavities coated inside with diffuse black and mounted next to each other on the same heat sink. The flat bottom of the cavities are in fact heat flux transducers on which heating elements have been mounted. Both cavities see the same thermal environment through accurately know circular apertures. A comparison of the power generated inside the cavities (very similarly as an househould weight can be used) is done. For instance a constant electrical power is generated in one of the channels and the difference between the two heatflux sensors is automatically brought back to zero by an ad hoc accurate servosystem that provides electrical power to the other channel called "active channel". This one is regularly irradiated by the Sun or closed. The difference of the electrical power fed to the active channel when its shutter is open (exposed to the Sun) and when it is closed is proportional to the incident solar irradiance. From time to time, the roles of the left channel and the right channel are reversed for half an hour with the purpose of monitoring the aging of the continuously exposed left channel. The sampling rate of the PMO6 instrument is 1 solar total irradiance / 2 minutes, for DIARAD 1 solar total irradiance / 3 minutes.

The ACRIM contains four cylindrical bays. Three of the bays house independent heat detectors, called pyrheliometers, which are independently shuttered, self calibrating, automatically controlled, and which are uniformly sensitive from the extreme UV to the far infrared. Each pyrheliometer consists of two cavities, and temperature differences between the two are used to determine the total solar flux. One cavity is maintained at a constant reference temperature, while the other is heated 0.5 K higher than the reference cavity and is exposed to the Sun periodically. When the shutter covering the second cavity is open, sunlight enters, creating an even greater difference in cavity temperatures. The power supplied to the second cavity by the ACRIM electronics decreases automatically to maintain the 0.5 K temperature difference between the two cavities. This decrease in the amount of electricity is proportional to the solar irradiance entering the cavity. Exposing the sensors to the space environment and the Solar UV radiation causes some small changes on the surface of the cavities which may affect the measurements. The ACRIM instrument monitors this type of problem by carrying three similar sensors, two of which are normally covered. At times these are opened for comparison purposes. Further details can be found in Wilson (1981, 1984).

Measuring the solar constant one finds:

- Part of the energy is blocked by dark sunspots and subsequently released in faculae. The screening effect by sunspots is overcompensated by the energy storage and release. This is demonstrated in Fig. 7.10.

- There are variations of the solar constant with the solar cycle.

First measurements with the ACRIM 1 (Active Cavity Radiometer Irradiance Monitor) experiment on board the Solar Maximum Mission and the ERB experiment on the Nimbus-7 satellite showed a positive correlation between the solar cycle activity, measured by the sunspot index, and the total solar irradiance. The peak to peak variation of about 1 W/m^2 (out of about 1367) between solar maximum and minimum was reported by Fröhlich (1987), Willson and Hudson (1988)

7.4. THE ATMOSPHERE'S RESPONSE TO SOLAR IRRADIATION

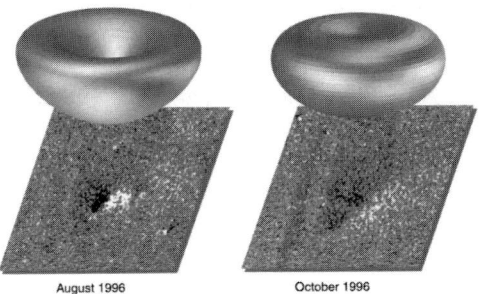

Figure 7.10: Three-dimensional rendering of the angular distribution of the excess irradiance emitted at 500 nm by the active region studied at two stages of its development, together with the magnetogram. A more uniform brightening of the facular region at the later stage is apparent (after Vicente Domingo).

and Foukal and Lean (1988). Somewhat larger fluctuation up to 0.2% occur over timescales of days and weeks.

Given that the total variation between the peaks of solar cycles 21 and 22 was about 0.1%, how much is the effect to be expected for a change of the corresponding global temperature on Earth? It is expected that this change of the solar irradiation produces a corresponding variation of about 0.2^0 C in globally averaged equilibrium surface temperature (Hansen and Lacis, 1990). But there is some considerable delay in the response. Because of the thermal inertia of the oceans, the time needed to approach equilibrium is much longer than 11 years (e.g., Reid, 1991), so that the actual temperature response to the observed variation during a solar cycle is likely to be considerably smaller, and probably insignificant from a climatic point of view.

In order to study a long term variation of the solar output, there is no direct observational support. It is therefore necessary to use proxy data or solar activity indicators. Sunspot index measurements exist over a time span of roughly 350 years and they suggest the presence of a 76-80 yr cycle, the Gleissberg cycle, modulating the 11 yr cycle (Sonett, 1982, Berry, 1987). Foukal and Lean (1990) gave an empirical model of total solar irradiance variation between 1874 and 1988.

The presence of the solar cycle has been claimed in various sets of proxies:

- auroral activity,

- isotopic composition of ice cores (Johnsen et al. 1970),

- tree growth, dendroclimatic investigations (Svenonius and Olausson, 1979). For annual rings to form, trees must "shut down" growth at some point to form a distinct ring boundary. This occurs in the dormant season, usually in the fall and winter. In the tropics, the seasons are not as distinct, so that trees can grow year-round. One fundamental principle of dendrochronology

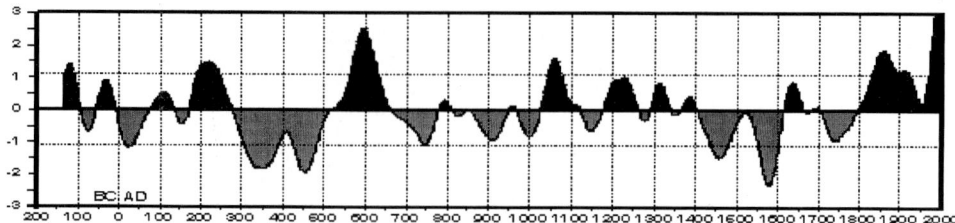

Figure 7.11: Reconstructed precipitation in northern New Mexico. Courtesy: Henri D Grissino-Mayer

is "the present is the key to the past," originally stated by James Hutton in 1785. However, dendrochronology adds a new "twist" to this principle: "the past is the key to the future." In other words, by knowing environmental conditions that operated in the past (by analyzing such conditions in tree rings), we can better predict and/or manage such environmental conditions in the future. Hence, by knowing what the climate-tree growth relationship is in the 20th century, we can reconstruct climate from tree rings well before weather records were ever kept! Let us give one example from Grissino-Mayer: Fig 7.11 shows a long-term precipitation reconstruction for northern New Mexico based on tree rings. How this reconstruction was made? The reconstruction was developed by calibrating the widths of tree rings from the 1900s with rainfall records from the 1900s. Because we assume that conditions must have been similar in the past, we can then use the widths of tree rings as a proxy (or substitute) for actual rainfall amounts prior to the historical record.

Individual tree-growth series can be "decomposed" into an aggregate of environmental factors:

$$R_t = A_t + C_t + \Delta D1_t + \Delta D2_t + E_t \qquad (7.20)$$

R_t is the tree ring growth as a function of t. A_t age related growth trend due to normal physiological aging processes, the climate (C) that occurred during that year the occurrence of disturbance factors within the forest stand (for example, a blow down of trees), indicated by $D1$, the occurrence of disturbance factors from outside the forest stand (for example, an insect outbreak that defoliates the trees, causing growth reduction), indicated by $D2$, and random (error) processes E not accounted for.

A study of tree rings and application to reconstruct climate was given by Cook et al. (1997). Sampling 300-to-500-year-old Siberian pine trees in the Tarvagatay Mountains of western central Mongolia, D'Arrigo et al.(1993) analyzed annual growth rings, which generally grow wider during warm periods and narrower in colder times in trees at the timber line. They developed a tree-ring record reflecting annual temperatures in the region dating back to 1550. The Mongolian tree rings show temperature changes that are strik-

7.4. THE ATMOSPHERE'S RESPONSE TO SOLAR IRRADIATION

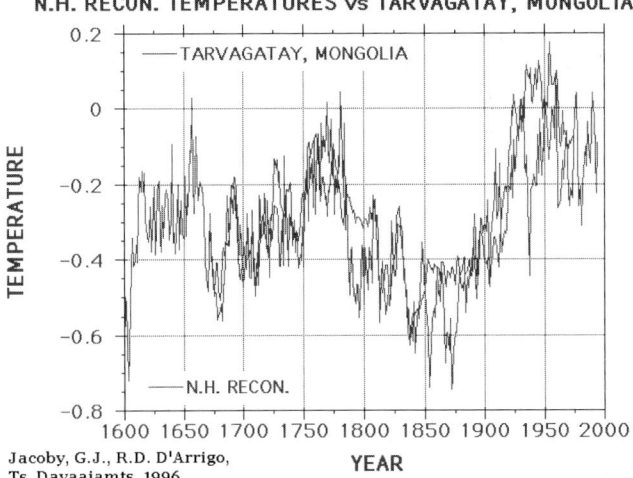

Figure 7.12: Temperatures derived from tree rings. Here the Maunder minimum is not seen whereas a cold period between 1830 to 1870)

ingly similar to records from tree rings in North America, Europe and western Russia. The general trends reflected in the tree-ring record include cooler conditions in the early 1700s, followed by warming that started mid-century. An abrupt cooling occurred in the late 1700s and continued for much of the 1800s. The coldest period was between 1830s and 1870s, after which a steadily increasing warming trend began. An example of this analysis is given in Fig 7.12.

- Solar radius variations (Gilliland, 1981),

- sedimentary rocks (Sonett and Williams, 1985)

- sea surface temperatures (Gerard, 1990). The mechanism how this could be related to solar irradiance variations works as follows:

  1. absorption of solar energy by the tropical oceans in a deep surface layer,
  2. transport of that energy by ocean currents,
  3. transfer of that energy by evaporation into atmospheric moisture and pressure systems leading to more precipitation (Perry, 1994).

Lewis et al. (1990) showed that solar radiation in visible frequencies, usually assumed to be absorbed at the sea surface, penetrates to a significant depth below the upper mixed layer of the ocean that interacts directly with the atmosphere. In clear water, the blue wavelengths, where the greatest amount of energy is available, penetrate the deepest, to nearly 100 m. Energy injected into the ocean at this depth can be stored for a substantial period of time.

As it has been stated above the transparency of the tropical oceans is dependent upon the amount of biogenic material, phytoplankton pigments, and degradation products that are present. In the Pacific Ocean, transparency increases from east to west, with the greatest penetration of solar energy occurring in the western tropical Pacific. Due to ocean currents, the North Pacific Ocean takes approximately 4 years to move temperature anomalies from the western tropical Pacific to near North America (Favorite and McLain, 1973).

During the prolonged period between 1500 and 1850, average temperatures in Northern Europe were much colder than they are today, this is known as the little Ice Age. The coldest part of this period coincides with a conspicuous absence of sunspots and other signs of solar activity, called the Maunder Minimum.

For example Gilliland reported a 76 year cycle in the solar radius, inferred from a 258 year record of transits of the planet Mercury, solar eclipse records and meridian transit measurements. Ribes et al. (1987) also reported as Gilliland that the solar radius is slightly increased in times of low solar activity during the Maunder minimum.

A review book on the role of the Sun in climate change was written by Hoyt and Schatten (1997) where other references can be found.

UV Radiation

Solar radiation shortward 320 nm represents only 2% of the total solar irradiance; 0.01% of the incident flux is absorbed in the thermosphere at about 80 km and 0.2% in the stratosphere above 50 km. This radiation is extremely important since the thermal structure and photochemical processes above the troposphere are controlled by it. The stratosphere is controlled by absorption and dissociation of O_2 in the 175 to 240 nm range. The 205 to 295 nm range is predominantly absorbed by ozone O_3. If there is a stratosphere- troposphere coupling, this could affect also the climate. The short term variation of UV radiation is ascribed to the evolution and rotation of plage regions on the solar disk. The XUV induced thermospheric temperature changes is shown for low and high solar activity in Fig. 7.13. Solar activity is measured in terms of the 10.7 cm radio flux and of the plage area A_p.

Energetic particles

There are three main contributions:

- electrons: they reach the high latitude thermosphere after interaction with the geomagnetic field and acceleration;

- high energy solar protons: their flux is enhanced during periods of large flares;

- galactic cosmic rays: they originate from outside the heliosphere but their input on Earth is partly controlled by solar activity.

7.4. THE ATMOSPHERE'S RESPONSE TO SOLAR IRRADIATION

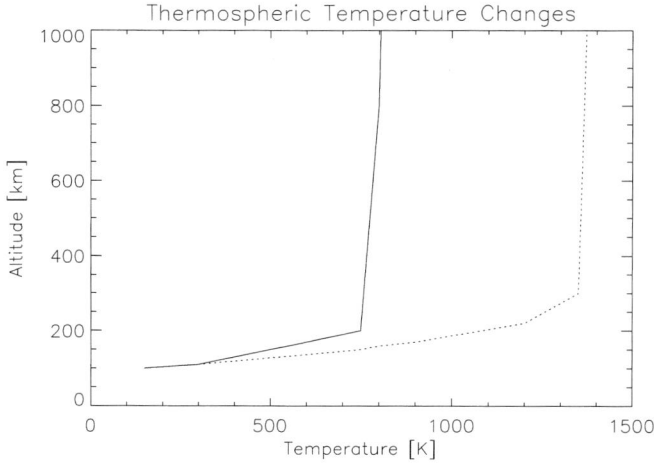

Figure 7.13: Thermospheric temperature changes, a) low solar activity, $F_{10.7} = 80, A_p = 0$, b) high solar activity, $F_{10.7} = 200, A_p = 80$

During large flares, intense fluxes of energetic protons ($10...10^4$ MeV) penetrate the Earth's polar cap regions. They produce ionization between 100 and 20 km. Such an event can last for a few hours to a few days. Large numbers of NO_x molecules are produced leading to a subsequent ozone depletion.

Relativistic electron precipitation are possible sources for ionization and odd nitrogen production at altitudes above 80 km, thus well above the ozone layer.

Pre Main Sequence Sun

All calculations show that during the early life of the Sun, the UV flux was much higher than today. The Sun had a behavior similar to a T Tauri star. Zahnle and Walker (1982) calculated that the flux decreases as

$$\sim t^{-s} \quad 0.5 < s < 1 \tag{7.21}$$

The exponent in this formula depends on the wavelength considered. Similar results were obtained by Canuto et al. (1982).

7.4.3 Response of the Earth's Atmosphere

In order to study the influence of varying solar irradiation to the Earth's atmosphere we must make a distinction between the different layers of the atmosphere and the wavelength of the solar radiation. As we have mentioned above, the penetration of solar radiation strongly depends on its wavelength, the larger the wavelength the deeper the penetration (see Fig. 7.14).

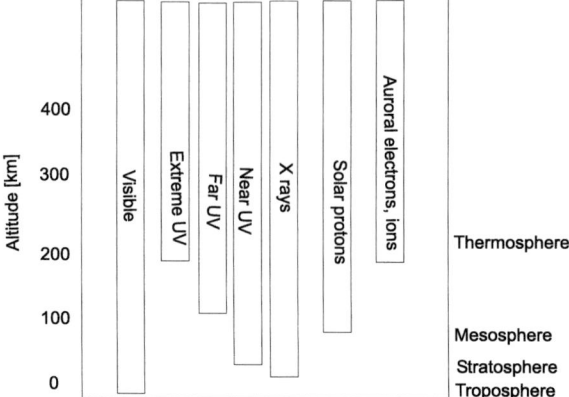

Figure 7.14: Penetration of different solar light waves resp. their induced particles in the atmosphere

Thermosphere and Exosphere

The thermosphere starts at a height of about 90 km and ends at about 250 km at the so called exobase. In the thermosphere the temperature gradient is positive, in the exosphere collisions become negligible, particles execute ballistic orbits. The most important heat source is solar XUV radiation creating the ionosphere. The resulting heating is conducted down to the mesopause where it can be radiated. The exosphere is approximately isothermal because it lies above the level where most of the energy is deposited. Also the thermal time constant is very short because of the low density there.

The variations can be divided into diurnal variations and longer term variations.

- The diurnal variations show a day/night ratio of 1.28 over the equator with the peak occurring about 2 p.m. During the night, heat is conducted down from the top of the thermosphere to its base, the mesopause, where it is radiated.

  The effect of the sun can be expressed and measured by the 10.7 cm radio flux which is given in the units 10^{-22} Wm^{-2}Hz^{-1}. If this quantity is multiplied by a factor of 1.8 deg per unit of flux, we obtain the temperature.

- The other variation comes from the solar activity cycle (see Table 7.7).

The thermospheric temperature changes are illustrated in Fig. 7.13. Similarly the peak electron density varies by a factor of 2 which is very important for shortwave radio communication. The temperature changes are obtained from the 10.7 index with a multiplier of 3.6. This value is twice than that for the 27 day variation which is due to solar rotation. The difference may result from the fact that in the case of activity cycle variations not only the variation of XUV but also a

7.4. THE ATMOSPHERE'S RESPONSE TO SOLAR IRRADIATION

Table 7.7: Exospheric temperature at solar maximum and minimum

	Temperature of exosphere
solar minimum	700 K
solar maximum	1200-1500 K

contribution from auroral heating- which is triggered by the solar wind- must be taken into account.

Finally we have to stress that both UV and particle precipitation have chemical effects and the most important is the production of N, NO and NO_2 (which is collectively called NO_x). The following reactions define the NO_x production in the thermosphere:

$$O^+ + N_2 \rightarrow NO^+ + N \qquad (7.22)$$
$$N_2^+ + O \rightarrow NO^+ + N \qquad (7.23)$$
$$N_2^+ + e \rightarrow N + N \qquad (7.24)$$
$$NO^+ + e \rightarrow N + O \qquad (7.25)$$

Mesosphere and Stratosphere

This region extends from the tropopause to the mesopause, at approximately 90 km. It is in local radiative equilibrium except from heat flowing in from the thermosphere. The radiative heating is by absorption of planetary radiation mainly by:

- CO_2 at 15 μm,
- O_3 at 6.3 μm
- absorption of solar UV by O_3.

The CO_2 band is the principal radiator. The O_3/CO_2 ratio decreases upwards; thus the heating to cooling also decreases and the temperature gradient is negative. In the stratosphere ozone begins to be more and more attenuated, a temperature maximum at 50 km occurs, the stratopause.

Which term in the heat balance of these atmospheric layers depends on the solar activity? It is the UV flux which is not variable at large scales thus the temperature changes to be expected from a variation of that flux should also be small. Most effects therefore come from ionization and a changing chemistry. The changes of the UV flux from the sun have only a modest effect on ozone amounts because both production (by a photolysis of O_3) and destruction are affected in the same way. Another effect is the penetration of solar protons or relativistic electrons into the middle atmosphere. By that penetration considerable amounts of NO_x are produced; these enhancements of NO_x increase the destruction of ozone at high altitudes. This could explain the inverse correlation of ozone amounts with solar activity found by Ruderman and Chamberlain (1975).

To test these predictions it is important to have data at the time scale of the solar cycle; however we must also take into account the instrumental drifts as well as the typical lifetime of the instruments which normally are below 5 yr. In the stratosphere, the ozone response is caused primarily by changes in production from O_2 and has a maximum value of 0.5 % for a 1 % change in the UV at 205 nm. The study of the response of the temperature has been made by Hood (1986, 1987b) and Keating et al. (1987) between 30 and 0.2 mbar (24 to 60 km) and later by Clancy and Rusch (1989) up to 90 km. They establish the already mentioned 0.5 % response. The very small temperature response lags the UV by 4 to 14 days. A study by Angell and Korshover (1975, 1976) established a correlation of the ozone column with solar activity with a peak to peak variation up to 10 % at 70^0 latitude and only 4 % at 47^0.

There seems to be no correlation of polar stratospheric temperatures and solar activity (Labitzke (1987), Labitzke, Van Loon (1988), Kerr (1988). There exists a stratospheric biennial oscillation which is more or less periodic and reversal of winds in the lower equatorial stratosphere with an average period of 27 months.

7.4.4 Troposphere

As we have seen above, only wavelengths below 300 nm penetrate to the troposphere and surface. We have already stressed that this part of the solar spectrum is only slightly variable with a peak to peak variation of about 1 part in 1400. Thus the troposphere which contains 90 % of the total mass of the Earth's atmosphere is subject to a nearly constant driving solar energy.

However, there have been innumerable attempts to find correlations between solar activity and various meteorological phenomena and other variables. If the troposphere is to be significantly influenced by the tiny changes of solar irradiation, there should exist a very strong mechanism of amplification (trigger mechanism). Such mechanisms were discussed:

- magnetospheric effects by electric field - including also effects of thunderstorms (Mc Cormac and Seliga, 1979)

- Hines (1974) suggested a change of the transmissivity of the stratosphere to upwardly propagating atmospheric waves (Callis et al. 1985 showed from models that this is possibly not the case)

- The effect found by Labitzke (1987): temperatures in the polar winter are jointly influenced by the solar cycle and the quasi biennial oscillation and the effect on the troposphere is discussed in Van Loon and Labitzke (1988).

- Eddy (1976, 1988) discussed the absence of sunspot activity during the 17 th century which is known as the Maunder minimum and an earlier event, called the Spörer minimum. Both periods seem to coincide with periods of reduced global temperatures the more recent is called the Little Ice Age. Eddy (1988) showed that the required solar input reduction would have to be much greater than the tiny amplitudes detected on the time scale of a solar cycle. Maybe also amplifying factors have to be considered.

7.4.5 Long Term Changes in Solar Irradiance

Indicators of solar activity such as ^{14}C concentration measurements and of climate (e.g. glaciers) show a clear correlation over the last 7 000 years. This was shown by Eddy in 1977. Considering a time series since 1860 the solar cycle length shows an excellent correlation with northern hemisphere land temperatures (Friis-Christensen and Lassen, 1991). For recent data however, these two parameters diverge. If there exists a global climate contribution of solar irradiance variations then there are three possible interactions or couplings between these variations and the Earth's climate:

- Variations of the total solar irradiance.

- Variations of the Sun's spectral irradiance; this denotes changes in the luminosity of the Sun in a given wavelength range. As we have discussed above, UV radiation influences atmospheric chemistry (production or destruction of ozone, see also Haigh 1994, 1996).

- Variations in the heliospheric magnetic field which are coupled to changes in the solar wind and influences the number and energy spectrum of cosmic ray particles. This was investigated by Potgieter, 1998 and Simpson, 1998. The variation of cosmic ray particles seems to be related to global cloud cover (see Svensmark and Friis-Christensen (1997) or Svensmark (1998) or Marsch and Svensmark (2000).

As we have mentioned above, the total solar irradiance varies by 0.1% and these measurements have been made very accurately since 1978 (a review about that was given by Fröhlich, 2000) The cycles covered by these measurements are 21, 22 and 23 (only 22 full). Of course from these time series it is impossible to extrapolate to earlier time series when the Sun was more active (e.g. cycle 19) or less active. One further problem of the time series available is that with the exception of SOHO/VIRGO they are restricted to the UV.

Irradiance variations of the past solar cycles can be determined from the surface distribution of the magnetic field if records of the field distribution or of proxies are available. The following proxies can be used:

- relative sunspot number (since 1700),

- group sunspot number (since 1610),

- sunspot and facular areas (A_s, A_f, since 1874,)

- Ca II plage areas (A_p, since 1915).

Using these data, one can reconstruct the cyclic component of the irradiance back to the Maunder minimum.

As a large sunspot group passes across the solar surface, there is a dip in the total solar irradiance. The variation is in the range of 0.2 promille.

Lockwood et al. (1999) reconstructed the aa-index of geomagnetic activity and found that the interplanetary magnetic flux at minimum of solar activity (that can be be reconstructed using the aa-index) has roughly doubled since 1900. This is in good agreement with ^{10}Be concentration in Greenland ice (Beer, 2000). ^{10}Be is produced by the interaction of cosmic rays with constituents of the Earth's atmosphere. The cosmic ray flux is modulated by the heliospheric magnetic field.

Lean et al. (1995) assumed that the background irradiance is proportional to the amplitude of the solar cycle; Hoyt and Schatten (1993) propose a trend corresponding to cycle length and Baliunas and Soon (1995) demonstrated that the amplitudes of stellar cycles (observed in Ca II H and K) scale with the length of the stellar cycle.

A short overview of long term chances in solar irradiance was given by Solanki and Fligge (2000).

The question whether the Earth's climate is influenced by solar activity has a central position in the present debate about the global warming. Greenhouse gas concentrations have a continuous increase and do not follow the observed decrease in the 1900's and in 1940-1970 example. These variations might be better explained when solar activity is taken into account. During a normal sunspot cycle the irradiance changes by 0.1% but could be greater (e.g. during the Maunder Minimum 0.3%, Lean (1997)).

7.4.6 Solar Protons

Solar protons when hitting the atmosphere, break up molecules of N_2 and H_2O vapor. When nitrogen gas molecules split apart, they can create molecules, called nitrogen oxides NO, which can last several weeks to months depending on where they end up in the atmosphere. Once formed, the nitrogen oxides react quickly with ozone and reduce its amounts. When atmospheric winds blow them down into the middle stratosphere, they can stay there for months, and continue to keep ozone at a reduced level.

Similarly water vapor molecules are affected by solar protons, breaking them up into radicals where they react with ozone. However, these molecules, called hydrogen oxides, only last during the time period of the solar proton event. These short-term effects of hydrogen oxides can destroy up to 70 percent of the ozone in the middle mesosphere. At the same time, longer-term ozone loss caused by nitrogen oxides destroys a maximum of about nine percent of the ozone in the upper stratosphere. Only a few percent of total ozone is in the mesosphere and upper stratosphere with over 80 percent in the middle and lower stratosphere.

The impacts on humans are minimal. NASA's HALOE was launched on the UARS spacecraft September 15, 1991 as part of the Earth Science Enterprise Program. Its mission includes improvement of understanding stratospheric ozone depletion by measuring vertical profiles of ozone, hydrogen chloride, hydrogen fluoride, methane, water vapor, nitric oxide, nitrogen dioxide, aerosols, and temperature. The SBUV/2 instrument was launched aboard the NOAA-14 satellite on December 30, 1994 and its mission is to observe the ozone layer.

7.5 Cosmic Rays

Galactic cosmic rays ($CGRs$) are high energy particles. They flow into our solar system from far away consisting mainly of electrons, protons and fully ionized atomic nuclei. These particles have speeds near the speed of light.

Since the particles are charged, they are deflected by the magnetic field of our Galaxy and the heliosphere as well as the Earth's magnetic field. Thus we can no longer point back to their sources in the Galaxy. A map of the sky with cosmic ray intensities would be completely uniform.

One of the direct observations we can make is to analyze the composition of the GCR's which tells us something about the origin. One finds all natural elements in roughly the same proportion as they occur in the solar system. However a detailed analysis of the differences in the abundances can give hints to the origin of the particles. Two properties of the particles can be measured:

- Determine which element; this is very easy since the different charges of each nucleus give different signatures.

- Isotopic composition; to determine the isotopic composition which in some cases gives better insights in the origin of the particles, the atomic nuclei have to be weighted which is much more difficult.

90% of the cosmic ray nuclei are protons, 9% alpha particles (ionized He) and the rest of the elements makes up only 1%. In this small percentage some elements are very rare and therefore large detectors are required.

Most cosmic rays are accelerated in the blast waves of supernova remnants. Remnants of supernova explosions are very active, we observe expanding clouds of gas, magnetic field activities etc. which can accelerate particles. Such processes in supernova remnants can last for several 10^3 yrs. The particles are accelerated in the magnetic field until they have enough energy to escape and become cosmic rays. Thus they can only be accelerated up to certain maximum energy which depends upon the size of the acceleration region and the magnetic field strength.

The problem we face is that cosmic rays have been observed at much higher energies than those supernova remnants can generate. Possible explanations for such extreme high energetic particles are:

- their nature is extragalactic: from galaxies with very active galactic nuclei,

- they are related to the gamma ray bursts,

- they are related to exotic particles which are predicted by several physical theories concerning with the origin of the universe; superstrings, exotic matter, strongly interacting neutrinos,

- they are topological defects in the very structure of the universe.

As we have stated above, cosmic rays include a number of radioactive nuclei whose numbers decrease through the radioactive decay. Measurements of these nuclei can be used therefore (as in the C^{14} method) to determine how long it has been since cosmic ray material was synthesized in the galactic magnetic field.

Let us now describe the interaction of cosmic ray particles with the heliosphere. The heliosphere is defined by the interplanetary magnetic field. This shields the interstellar plasma (charged particles). Interstellar neutral gas flows through the solar system however, since uncharged particles are not influenced by magnetic fields. The speed is approximately 25 km/s. When approaching the Sun, these neutral atoms become ionized by two processes:

- photo-ionization: an electron of the neutral atom is knocked off by a solar high energy photon (e.g. a UV photon);

- charge exchange: an electron is exchanged to an ionized atom of solar wind particle.

As soon as these particles are charged the Sun's magnetic field carries them outward to the solar wind termination shock region. The ions repeatedly collide with the termination shock, gaining energy during each collision. This continues until they escape from the shock region and diffuse back toward the inner heliosphere. Such particles are called anomalous cosmic rays (ACRs). ACRs are thought to originate from the very local interstellar medium and are not related to the above mentioned violent processes as the GCRs. They can easily be discerned from GCRs because they have lower speed and energy. They include large quantities of He, O, Ne and other elements which have in common high ionization potentials.

The third component of cosmic ray particles are Solar energetic particles (*SEP*). They move away from the Sun due to plasma heating, acceleration and other processes. Flares e.g. inject large amounts of energetic nuclei into space, the composition varies from flare to flare. On the scale of cosmic radiation, SEPs have relatively low energies.

7.5.1 The Heliosphere

The heliosphere as already stated, is the magnetic shield caused by the Sun which protects us against energetic cosmic ray particles. The solar wind which is a continuous stream of plasma expands out through the solar system until it changes from supersonic to subsonic speed what is called a termination shock. The distance of that region is assumed to be about 200 AU and the space within is called the Heliosphere which encloses the whole solar system (e.g. Pluto's mean distance is only 39 AU). Because of the magnetic fields, only some of the GCR particles penetrate to the inner part of the solar system. Thus the magnetic field of the heliosphere works as a shield. The magnetic activity of the Sun changes however with the solar cycle (every 11 year the Sun's magnetic field reverses, the true cycle is thus 22 years). This causes a variation of the GCR flux. When the Sun is more active, the magnetic field is stronger, and as a result, fewer GCR arrive in the vicinity of the Earth. We can also say that the higher the energy the particles have, the less they are modulated by the solar cycle.

Instrumental recordings of cosmic rays started around 1935. The first measurements where done with ionization chambers, which measure mainly the muon flux. These muons are responsible for most of the ionization in the lower part of

7.5. COSMIC RAYS

the troposphere. With a neutron monitor one can measure the low energy part in the GCR spectrum.

For our study here, it is important to note that by measuring cosmic rays one can derive a proxy for solar activity very long back in time. This is possible since isotopes in the atmosphere are produced by cosmic rays. From such recordings a good qualitative agreement between cold and warm climatic periods and low and high solar activity during the last 10 000 years was found. When we consider ^{14}C variations during the last millennium, one can deduce, that from 1000-1300 AC solar activity was very high which coincided with the warm medieval period. We know from history that e.g. during that period the Vikings settled in Greenland. The solar activity - if it is well represented by the ^{14}C variation- decreased and a long period followed which is now called the little ice age (in this period falls also the so called Maunder Minimum, 1645-1715, where practically no sunspots were observed). This period lasted until the middle of the 19th century. From then on, solar activity has increased and is the highest in the last 600 years.

Thus we may assume the following connections:

$$\text{low solar activity} \rightarrow \text{weak magnetic field} \rightarrow \text{more GCRs} \rightarrow \text{more}\ ^{14}C \quad (7.26)$$

If that assumption is true, there is a mechanism, how the Earth's climate can be influenced by the Sun.

7.5.2 Cloud, and Cloud Formation Processes

Clouds are created by condensation or deposition of water above the Earth's surface when the air mass becomes saturated (relative humidity 100 %). Saturation can occur when the temperature of an air mass goes to its dew point or frost point. There are different mechanisms to achieve this:

Orographic precipitation: this occurs when air is forced to rise because of the physical presence of elevation (land). As such a parcel of air rises it cools due to the adiabatic expansion at a rate of approximately 10 degrees per 1 000 m. The rise of the parcel is stopped if saturation is reached. An example of this mechanism is the west coast of Canada with large precipitation.

Convectional precipitation: this is associated with heating of the air at the Earth's surface. When there is enough heating, the air becomes lighter than the surrounding masses, begins to rise (cf. a hot air balloon begins to rise), expands and cools as above. When sufficient cooling takes place, saturation is reached again forming precipitation. This mechanism is active in the interior of continents and near the equator forming cumulus clouds and thunderstorms.

Convergence or frontal precipitation: this mechanism takes place when two masses of air come together. One is usually moist and warm and the other is cold and dry. The leading edge of the cold front acts as an inclined wall or front causing the moist warm air to be lifted. Then the above described processes start again: rise, cooling and saturation. This type of precipitation is common in the mid latitudes.

Finally we have to mention the radiative cooling: this occurs when the Sun is no longer supplying the ground and overlying air with energy due to insolation

Table 7.8: Various influences on the climate

S, solar constant (at 1 AU)	1360 W/m^2
S/4, top of atmosphere	340 W/m^2
S/4(1-a), a=0.3 Earth's albedo	235 W/m^2
1 % change in the albedo	1 W/m^2
estimated rad. effect due to CO_2 increase since 1750	1.5 W/m^2
doubling of CO_2	4 W/m^2
radiative effect of clouds (cooling)	17-35 W/m^2

during nighttime. The surface of the Earth begins to lose energy in the form of longwave radiation. This causes the ground and the air above it to cool. The precipitation that results from this kind of mechanism takes the form of dew, frost or fog.

Of course these mechanisms may act as a combination: convection and orographic uplift can cause summer afternoon showers in the mountains.

Let us compare the levels of cloud cover for summer and winter (northern hemisphere). For summer in the northern hemisphere, highest levels of cloud cover occur over the mid-latitude cyclone storm tracks of both hemispheres, Intertropical Convergence Zone over land surfaces, and the Indian Monsoon region (orographic lifting). Lowest values occur over the subtropical deserts, the subsidence regions of the subtropical oceans, and the polar regions. For winter in the northern hemisphere highest levels of cloud cover occur over the mid-latitude cyclone storm tracks of both hemispheres and the Intertropical Convergence Zone over land surfaces. Lowest values occur over the subtropical deserts, the subsidence regions of the subtropical oceans, and over the South Pole.

Clouds influence vertically integrated radiative properties of the atmosphere. They cause a cooling through reflection of incoming shortwave radiation (sun - light) and heating by absorption and trapping of outgoing long wave radiation (thermal radiation). Let us consider the net radiative impact of a cloud: this mainly depends on two parameters, on its height above the surface and its optical thickness. High optically thin clouds tend to heat while low optically thick clouds tend to cool. The net forcing of the global cloud cover is in the range between $17 - 35$ W/m^2, as it is derived from climate models. Thus a significant influence on the global cloud cover can be potentially very important for Earth's climate (see also Table 7.8).

It has been found that the Earth's cloud cover follows the variation in GCR. It seems to be that the ionization in the atmosphere produced by GCR is the essential link. One can estimate that a variation in cloud cover of 3 % during an average 11-year solar cycle could have an effect of 0.8-1.7 W/m^2. This is a very significant amount.

The idea that cosmic rays can influence cloud formation was first pointed out by Svensmark (1997). He showed that there was a significant correlation between total cloud cover over the Earth and the influx of cosmic rays. The rays ionize

7.5. COSMIC RAYS

particles in the low troposphere which then seed the growth of cloud water droplets. During the past century the shielding from cosmic rays has increased since solar activity has increased. This decreases the formation and cooling influence of low clouds and may thus provide a possible contribution to the global warming of the past 100 years (Marsh, Svensmark, 2000; Svensmark, 1999).

Let us consider and summarize the changes in the magnetic field in the solar atmosphere. Shorter solar cycles facilitate a rise in the coronal source flux, longer cycles allow it to decay. The accumulation of the coronal source flux strongly depends on the rate of flux emergence in active regions. In general the peak and cycle averaged sunspot numbers are larger when cycles are shorter. Therefore, shorter cycles are associated with larger flux emergence rates, there is less time for the open flux to decay. We can state:

shorter activity cycle → increased coronal flux

The coronal source surface is where the magnetic field becomes approximately radial. This occurs at $r = 2.5 R_\odot$. This surface can also be regarded as the boundary that separates the solar corona from the heliosphere. The magnetic flux threading the corona source surface is called F_s or open solar flux. If there is a rise of the flux F_s than the cosmic ray flux incident on the Earth will decrease. Lockwood and Foster (2000) estimated that the cosmic ray flux > 3 GeV was 15% larger around 1900 than it is now. As it was shown above, cosmic rays generate air ions in the sub ionospheric gap which allows current to flow in the global electric current. This connects thunderclouds with the ground via lightening.

Table 7.9: Causes of Global Warming of about 0.5 C, 1880-1997

Climate Forcing Factor	Est. forcing ^{0}C , 1880-1997
Solar luminosity increase	+0.25[1]
Decrease in volcanic stratospheric aerosols	+0.15[2]
Increased anthropogenic sulfate aerosols	Up to -0.1 C
Increased anthropogenic carbon aerosols	Up to + 0.1 C (offsets sulfate aerosols)
Carbon dioxide warming	+0.05 to +0.10 C
Decrease in stratospheric ozone	-0.05 C[3]
Increase in cirrus contrails from airplanes	+ 0.05 C
Urban heat island effects	+0.01 to +0.10[4]
Changing skyline effects	possibly as large as + 0.25 C
Sum total of all above forcing factor	+0.51 to 0.60 C

[1] See "The Role of the Sun in Climate Change"; also see Lean et al., 1995 .
[2] Wu et al., 1990
[3] Schwartz and Andreae 1996
[4] Balling, 1992

An analysis of ISCCP D2 cloud data showed a correspondence between low cloud cover and cosmic ray flux (Palle and Butler, 2000). The authors also mentioned that the effect of increased global sea temperatures, increased aerosols and aircraft traffic on cloud formation processes should be taken into account.

7.6 What Causes the Global Warming?

This is a very strong debate. In the extreme case the warming is not caused by a substantial greenhouse effect since as is shown in Table 7.9 other factors can contribute to the observed increase in temperature. However, all these estimates are estimates and should be taken with caution.

A summary of the effects is illustrated in Fig. 7.15. The data shown in that Fig are i) Reconstructed NH temperature series from 1610-1980, updated with raw data from 1981-1995 ii) Greenhouse gases (GHG) represented by atmospheric CO_2 measurements (iii) Reconstructed solar irradiance (see Lean et al, 1995) (iv) Weighted volcanic dust veil index (DVI) (v) Evolving multivariate correlation of NH series with the 3 forcings (i) (ii) and (iii). The data are from Mann et al. (1999, and further references therein). These authors conclude that while the natural (solar and volcanic) forcings appear to be important factors governing the natural variations of temperatures in past centuries, only human greenhouse gas forcing alone, can statistically explain the unusual warmth of the past few decades.

7.6. WHAT CAUSES THE GLOBAL WARMING? 153

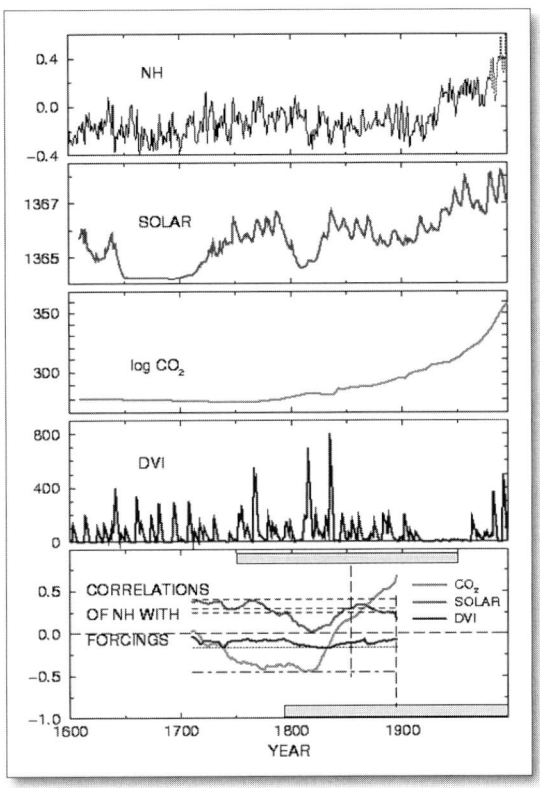

Figure 7.15: Relationship of Northern hemisphere mean (NH) temperature reconstruction to estimates of three candidate forcings between 1610 and 1995.

Chapter 8

Space Weather and Radiation Damage

In this chapter we discuss the influences of radiation damage both to humans in space as well as to electronics and solar panels of satellites.

8.1 The Early Sun

8.1.1 T Tauri stars

T Tauri stars are a group of stars which are solar like and often associated with molecular clouds. They are very early stars that means that they have not yet reached the main sequence and they are still contracting (see the introduction about stellar evolution). Their masses and temperatures are quite similar to the Sun but they are brighter. Their rotation rate is in the range of a few days (for the Sun it is about 1 month). They are active variable stars. The first ones were found about 1945 by their optical variability and chromospheric lines. Later on some evidence for large starspots on their surfaces were found. The X ray emission which is about 1 000 times that of the present Sun and radio flux is not constant. Some of them also show molecular outflow and strong stellar winds. By their IR and sub mm excess radiation it was found that about half of them are surrounded by circumstellar disks.

Contrary to normal main sequence stars like the Sun their energy is not produced by nuclear fusion near the core but by a slow gravitational contraction. T Tauri stars belong to the group of so called YSO (young stellar objects) of type II. Type I YSO are very young protostellar objects at the age of just a few 100 000 years. An example is HR 4796. At a wavelength of 12 μm the object appears as a point source, at a wavelength of 21 μm it is much larger and diffuse indicating a circumstellar disk of dust. They are bright in the mid and far IR but invisible in the optical.

8.1.2 The Early Sun

At an age of about 40 Million years our Sun became a zero age main sequence star. That means that it reached the main sequence and nuclear fusion of H to He started. At that time the Sun had about 70% of the total flux that is emitted presently. But in the UV and X-rays the flux was higher by a factor of about 100 than now. This of course has important consequences for the formation of the planets, their atmospheres etc. In its T Tauri and post T Tauri evolution the Sun's short wavelength emission was considerably higher than it is now. At that time the terrestrial planets were formed already, the protoplanetary disk evaporated, comets ejected out into the Oort cloud and the big bombardment period from the remaining rocky planetesimals and comets began; this caused probably several evaporations of the Earth's oceans.

How can we find indications for this T Tauri and Post T Tauri Phase of the Sun? Measurements of the ^{15}N to ^{14}N ratio in the atmosphere of the satellite Titan (which is Saturn's largest satellite) have shown that the bulk N is enhanced in the heavier ^{15}N isotope by about 4.5 times relative to the Earth's value. A $^{15}N/^{14}N$ anomaly on Mars of about 1.6 times the terrestrial value has also been found. These measurements can only be explained by the above mentioned T Tauri and post T Tauri phase of the early Sun (Lammer et al. 2000).

8.2 Radiation Damage on Living Organisms

8.2.1 Definitions

Radiation is energy in the form of waves or particles. X rays and gamma rays are electromagnetic waves of radiation, as is visible light. Particulate radiation includes alpha and beta radiation. The energy associated with any radiation can be transferred to matter. This transfer of energy can remove electrons from the atoms leading to the formation of ions. The types of radiation capable of producing ions in matter are collectively referred to as *ionizing radiation.*

Alpha particles are composed of two protons and two neutrons. Alpha particles do not travel very far from their radioactive source. They cannot pass through a piece of paper, clothes, or even the layer of dead cells which normally protects the skin. Because alpha particles cannot penetrate human skin they are not considered an *external exposure hazard* (this means that if the alpha particles stay outside the human body they cannot harm it). However, alpha particle sources located within the body may pose an "internal" health hazard if they are present in great enough quantities. The risk from indoor radon is due to inhaled alpha particle sources which irradiate lung tissue.

Beta particles are similar to electrons, except they come from the atomic nucleus and are not bound to any atom. Beta particles cannot travel very far from their radioactive source. For example, they can travel only about one half an inch in human tissue, and they may travel a few yards in air. They are not capable of penetrating something as thin as a book or a pad of paper.

Gamma rays are an example of electromagnetic radiation, as is visible light.

8.2. RADIATION DAMAGE ON LIVING ORGANISMS

Table 8.1: Radiation related units

Unit	Measures	Definition
Roengten (R)	exposure	1 R=2.56×10^{-4} C/s is deposited in dry air kg^{-1}; only for X rays
Radiation absorbed dose (rad)	absorbed dose	1 rad = absorption of 100 ergs per g material used for any type of radiation no description of biol. effects
Roengten equivalent man REM	equivalent dose	rem=rad $\times$ Q Q... quality factor (type of radiation) relates absorbed dose to effective biological damage
Gray, Gy	absorbed dose	1 Gy= 1J of energy deposed in 1 kg of material
Sievert Sv	equivalent dose	Sv = Gy $\times$ Q, Q...quality factor 1 Sv = 100 rem
Becquerel Bq	radioactivity	1Bq=1 transformation/sec 1 Cu=3.7×10^{10}Bq

Gamma rays originate from the nucleus of an atom. They are capable of travelling long distances through air and most other materials. Gamma rays require more "shielding" material, such as lead or steel, to reduce their numbers than is required for alpha and beta particles.

In Table 8.1 we give some definitions used in radiation physics.

The effect of radiation on any material is determined by the *dose* of radiation that material receives. Radiation dose is simply the quantity of radiation energy deposited in a material. There are several terms used in radiation protection to precisely describe the various aspects associated with the concept of dose and how radiation energy deposited in tissue affects humans.

Some terms related to radiation dose:

- Chronic dose: A chronic dose means a person received a radiation dose over a long period of time.

- Acute dose: An acute dose means a person received a radiation dose over a short period of time.

- Somatic effects are effects from some agent, like radiation that are seen in the individual who receives the agent.

- Genetic effects: Genetic effects are effects from some agent, that are seen in the offspring of the individual who received the agent. The agent must be encountered pre-conception.

- Teratogenic effects: Teratogenic effects are effects from some agent, that are

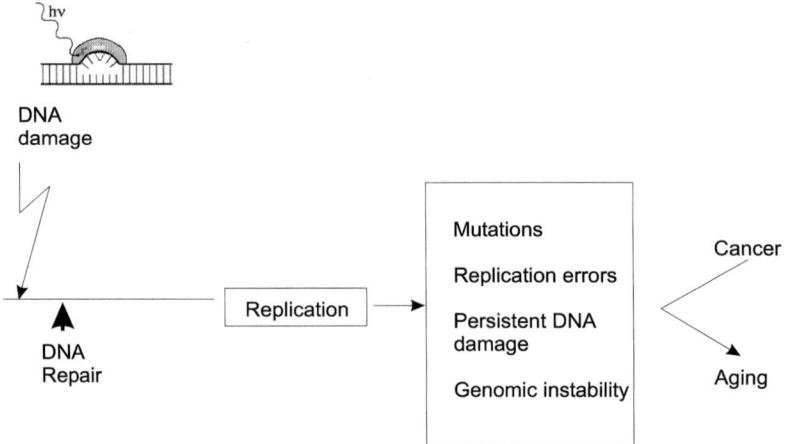

Figure 8.1: DNA damage caused by radiation

seen in the offspring of the individual who received the agent. The agent must be encountered during the gestation period.

8.2.2 Radiation Damage on DNA

The basic unit of any living organism is a cell. It is a small, watery compartment filled with chemicals and a complete copy of the organism's genome. The term genome denotes all the DNA in the cell (chromosomes and other). Different organisms have different numbers of chromosomes (e.g. humans have 23 pairs of chromosomes, 44 autosomes and 2 pairs of sex chromosomes). Each parent contributes one chromosome to each pair and so children get half of their chromosomes form their mother and half from their father.

The structural arrangement of DNA looks like a long ladder twisted into a helix. The sides of the ladder are formed by a backbone of sugar and phosphate molecules. The rungs consist of nucleotide bases joined weakly in the middle by hydrogen bounds. There are two major ways that radiation injures the DNA inside the cells of an organism:

- water in the body tends to absorb a large fraction of radiation and becomes ionized. When water is ionized it forms highly reactive molecules which are called free radicals. Those react with and damage the DNA molecules.

- radiation can also collide directly with the DNA molecules ionizing and damaging it directly.

The typical symptoms of radiation sickness are: severe burns that are slow to heal, sterilization, cancer. High doses are rapidly fatal (within days or weeks).

Yang et al. (1996) discussed DNA damage and repair in oncogenic transformation by heavy ion radiation. The most important late effect of energetic heavy

8.2. RADIATION DAMAGE ON LIVING ORGANISMS

Table 8.2: Radiation dose limits in mSv for astronauts

Time period	blood forming organs	eyes	skin
30 days	250	1000	1500
annual	500	2000	3000
career for males	2000 mSv+75(age[years]-30)	4000	6000
career for females	2000 mSv+75(age[years]-38)	4000	6000

Table 8.3: Total average annual radiation does in the US

radon in the air	2 mSv (56%)
rocks, building material	0.28 mSv (8%)
cosmic rays	0.28 mSv (8%)
natural radioactive material in body	0.39mSv (11%)
medical and dental rays	0.39 mSv (11%)
nuclear medicine tests	0.14 mSv (4%)

ions in cosmic rays and solar particle events is risk assessment in carcinogenesis.

8.2.3 DNA Repair

Wether or not a cell can repair depends on the damage to the DNA.

- single strand break in the DNA: this can be usually repaired and normal cell function is restored.

- breaks in both DNA strands: usually the damage is too severe to repair and the cell dies.

- chemical change or mutation: cannot be repaired; cancer or a mutation offspring results if this occurs in a sperm or egg cell.

8.2.4 Radiation Dose Limits for Astronauts

These limits were set by the US National Council on Radiation Protection and Measurements for all space missions in order to protect the astronauts. But there is an exception for exploratory missions and circumstances in space (e.g. mission to Mars). In Table 8.2 the relevant data are given.

The radiation dose limits for ordinary citizens are much lower. The annual dose is about 50 mSv, the lifetime dose is age [years] x 10 mSv. In the US the total average annual dose is about 3.6 mSv.

The single dose effects can be described as follows:

Table 8.4: Single dose effects

0.001 mSv	dental x rays
0.002 mSv	5 hr transcontinental flight
0.02 mSv	chest X-ray
1.000 mSv	radiation sickness
2500 mSv	sterility in females
3500 mSv	sterility in males
4000 mSv	average lethal dose (without any treatment)

Riklis et al. (1996) discussed biochemical radioprotection using antioxidants and DNA repair enhancement and found that the right combination proves effective in providing protection from a wide range of radiation exposures over a long period of time.

8.2.5 Genetic vs. Somatic Effects

Somatic effects of radiation damage appear on the exposed person. Prompt somatic effects appear after an acute dose. One example of a prompt effect is temporary hair loss. Delayed somatic effects may occur years after radiation doses are received. Typical effects are the development of cancer and cataracts. Let us briefly mention the most important syndromes.

- Blood forming organ (bone marrow) syndrome: damage to the cells which divide at the most rapid pace; bone marrow, spleen and limphathic tissue. Symptoms include internal bleeding, fatigue, bacterial infections and fever.

- Gastrointestinal tract syndrome (>1000 rad): damage to cells which divide less rapidly; lining to the stomach and intestines. Symptoms are nausea, vomitting, diarrhoea, dehydration, loss of digestion ability, bleeding ulcers.

- Central nervous system syndrome (> 5000 rad): damage to cells which do not reproduce such as nerve cells. Symptoms include loss of coordination, confusion, coma, shock.

It seems now that death is not caused by radiation damage on the nervous system but by internal bleeding and fluid and pressure build-up on the brain.

The genetic or heritable effect appears in the future generation of the exposed person as a result of radiation damage to the reproductive cells.

We have seen that satellite systems are vulnerable to Space Weather through its influence on energetic charged particle and plasma populations and that aircraft electronics and air crew are subjected to atmospheric secondary radiation produced by cosmic rays and solar particle events. This is discussed by Dyer (2001). The Advanced Composition Explorer (ACE) continuously monitors the solar wind and produces warnings by monitoring the high-energy particles that can produce radiation damage in satellite systems (Zwickl et al., 1998).

3.9-2.5 Billion years ago the Earth was dominated by an oceanic lithosphere. Cockell (2000) calculated that the DNA damage rates might have been approximately three orders of magnitude higher in the surface layer of the Archean oceans than on present-day oceans. However, at 30 m depth, damage might have been similar to the surface of present-day oceans. On the other hand, risk of being transported to the surface water in the mixed layer was quite high. Thus the mixed layer may have been inhabited by a low diversity UV-resistant biota. Repair capabilities similar to Deinococcus radiodurans would have been sufficient to survive in the mixed layer. During the early Proterozoic ozone concentrations increased and the UV stress would have been reduced and a greater diversity of organisms could have inhabited the mixed layer.

Lean (2000) discusses societal impacts of solar electromagnetic radiation.

The Yohkoh satellite was launched in 1991. Song and Cao (1999) discuss CCD radiation damage. Evans et al. (1999) discuss charged-particle induced radiation damage of a HPGe gamma-ray detector during spaceflight.

8.2.6 The Solar Proton Event in August 1972

Between the manned Apollo 16 and 17 missions one of the largest solar proton events ever recorded occurred. As a matter of luck no astronauts were in space during that time. Computer simulations were done later to reconstruct the influence on astronauts during that time. The main result of these simulations was that even inside of a spacecraft the astronauts would have absorbed a lethal dose of radiation within 10 hrs after the start of the event. At 6:20 UT an optical flare was observed on the Sun. At 13:00 UT the astronauts' allowable 30- day radiation exposure to skin and eyes was exceeded. At 14:00 the astronauts' allowable 30-day radiation exposure for blood forming organs and yearly limit for eyes was exceeded. The yearly limit for skin was exceeded at 15:00 UT. At 16:00 UT the yearly limit for blood forming organs and the career limit for eyes was exceeded. At 17:00 UT the career limit for skin was exceeded.

This event dramatically shows the need for space weather forecasting. The correlation of solar proton events with activity cycle is evident (Fig. 8.2.6).

Heckman (1988) discussed proton event predictions.

8.3 Solar UV Radiation Damage

8.3.1 General Remarks

Most UV radiation from the Sun is absorbed by the ozone layer or reflected back into space so only a small amount reaches the surface of the Earth. Sunlight is received as direct rays and as diffuse light, i.e. skylight which has been scattered by the atmosphere. The sky is blue because air molecules scatter the shorter wavelength (blue light) more than the red light, the index of scattering depends on the wavelength. UV light is scattered even more than blue light.

One has also to not forget the diffuse UV light and thus being shaded from direct Sunlight provides only a partial protection. Typical window glasses transmit

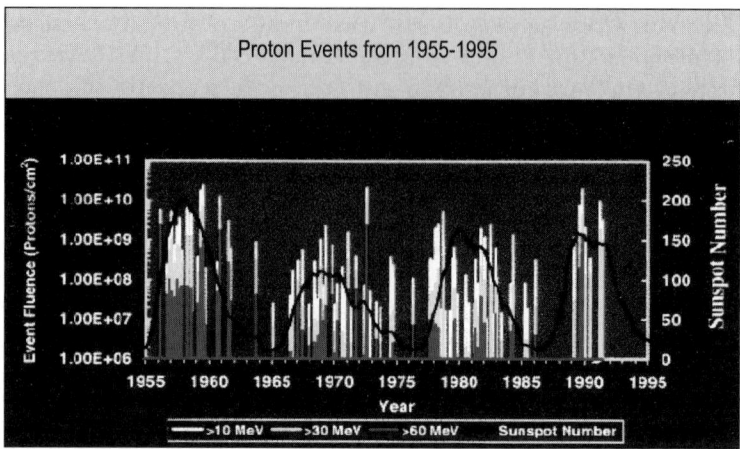

Figure 8.2: Correlation of the occurrence of solar proton events with solar activity cycle (indicated by the sunspot number)

less than 10% of ultraviolet light, and sunblock creams work by absorbing or reflecting UV rays. The SPF rating of sunscreens gives an indication of their effectiveness as UV blockers. For example, an SPF of 15 means that it should take 15 times as long to before skin damage occurs (i.e., the cream should block about 93% of the radiation that causes skin damage).

UV radiation is subdivided into three wavelength bands:

- UVA (315-400 nm), produces photochemical smog; damages plastic, paints and fabrics. UVA rays are not as energetic as UVB and, as a consequence, cause little sunburn or skin reddening. On the other hand, UVA rays penetrate deeper into the skin. The damage they cause is on a cellular level, occurring slowly and accumulating over a period of time. UVA radiation induces the formation of free radicals that, in turn, attack the lipids in the skin. The resulting damage gives rise to the visible signs of aging such as wrinkles and thickened skin. The skin's natural defenses against these free radicals are ascorbic acid (vitamin C) and alpha-tocopherol (vitamin E). These two vitamins are potent anti-oxidants that intercept the free radicals before they can do much damage. Vitamin C protects significantly better against UVA phototoxicity than vitamin E. Vitamin E, on the other hand, is more efficient against UVB.

- UVB (290-315 nm); 1% of solar radiation energy is in this band, most of it absorbed by ozone. Can damage DNA; smaller changes in ozone can lead to large changes in UVB radiation at the surface. Other effects are: Production of vitamin D in humans, skin cancer and damage to eye tissue. Plants and aquatic organisms suffer reduced growth, and many materials such as plastics degrade more rapidly in response to increased UVB radiation.

8.3. SOLAR UV RADIATION DAMAGE

- UVC (220-290 nm); totally blocked by ozone and other gases, does not reach the Earth's surface.

A person's potential to develop skin cancer is related to their exposure UVB radiation (sunburn). In New Zealand, about one person in three will develop a skin cancer during their lifetime. About half the number killed on the roads die of skin cancer in New Zealand. New Zealand and Australia have a very high melanoma incidence compared with other countries.

How can this be explained?

- New Zealanders have an outdoor lifestyle,

- wear fewer clothes now than in the past,

- the ancestors of most white-skinned New Zealanders migrated from the UK, which is at much higher latitude, and has much lower levels of UV radiation. These people are therefore poorly adapted to the relatively high levels of UV naturally present in New Zealand;

- calculations suggest that locations in the Southern Hemisphere should receive approximately 15% more UV than locations at a similar latitude north of the Equator (Basher, 1981; McKenzie, 1991). This is caused by differences in ozone between the Northern and Southern Hemispheres, and also because the Earth is slightly closer to the Sun during the Southern Hemisphere summer (McKenzie and Elwood, 1990);

- measurements show much larger differences, with biologically-damaging UV being 50-80% more in the Southern Hemisphere than at comparable Northern latitudes in Europe. The differences are caused by the buildup of tropospheric pollution (tropospheric ozone and aerosols) in the North (Seckmeyer and McKenzie, 1992);

- the largest levels of UV. Much higher levels of UV are experienced in countries, such as Australia, which are closer to the equator.

The amount of UVB light at ground level is determined by three factors: a) solar elevation, b) the amount of ozone in the atmosphere and c) the cloudiness of the sky. Please note that during local noon the amount of background radiation is the same as direct radiation three hours before and afterwards. At NZ's latitude, approximately 40% of the daily sunburn radiation occurs during the two hour period centered on solar noon.

Since the late 1970s an ozone hole has formed over Antarctica during early spring. The amount of ozone over New Zealand varies seasonally with a maximum in spring and a minimum in early autumn. Evidence of ozone destruction has also been observed over the Arctic. The ozone hole is caused by the special meteorological conditions of the cold atmosphere above polar regions which amplify the destructive ability of CFCs. The Antarctic ozone hole cannot shift over New Zealand. However, ozone losses over Antarctica may contribute to changes in ozone over the whole globe.

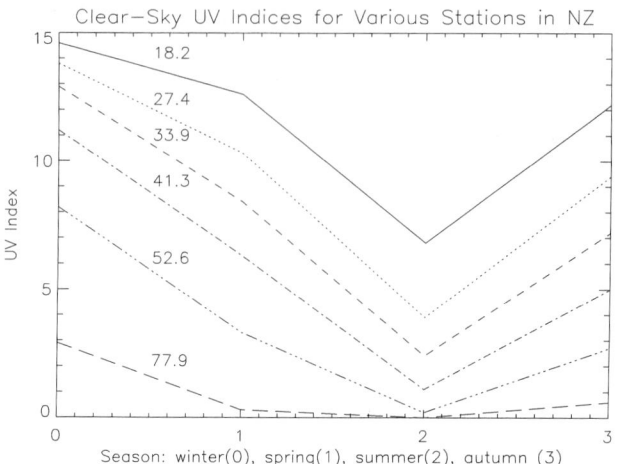

Figure 8.3: Typical clear-sky UV indices over New Zealand and its surrounding region. Seasonal variations are larger at low latitudes (denoted by numbers).

In Fig. 8.3 clear sky UV indices for different stations in NZ are given. Seasonal variations are higher at low latitudes as well as the absolute values.

8.3.2 Effects on the Skin

Generally, excessive UV exposure results in a number of chronic skin changes. These include:

- various skin cancers of which melanoma is the most life-threatening;

- an increased number of moles (benign abnormalities of melanocytes),

- a range of other alterations arising from UV damage to keratinocytes and blood vessels;

- UV damage to fibrous tissue is often described as "photoageing". Photoageing makes people look older because their skin loses its tightness and so sags or wrinkles.

United Nations Environment Programme (UNEP) has estimated that more than 2 million nonmelanoma skin cancers and 200 000 malignant melanomas occur globally each year. Let us assume that there is a 10% decrease of stratospheric ozone; then it is estimated that an additional 300 000 nonmelanoma and 4 500 melanoma skin cancers would result worldwide.

Caucasians have a higher risk of skin cancer because of the relative lack of skin pigmentation. The worldwide incidence of malignant melanoma continues to increase, and is strongly related to frequency of recreational exposure to the sun and to history of sunburn. There is evidence that risk of melanoma is also

8.3. SOLAR UV RADIATION DAMAGE

related to intermittent exposure to UV, especially in childhood, and to exposure to sunlamps. However, the latter results are still preliminary.

8.3.3 Effects on the Eye

The acute effects of UV on the eye include the development of photokeratitis and photoconjunctivitis, which are like sunburn of the delicate skin-like tissue on the surface of the eyeball (cornea) and eyelids. While painful, they are reversible, easily prevented by protective eyewear and have not been associated with any long-term damage.

Chronic effects however include the possible development of pterygium (a white or cream colored opaque growth attached to the cornea), squamous cell cancer of the conjunctiva (scaly or plate-like malignancy) and cataracts. Some 20 million people worldwide are currently blind as a result of cataracts. Of these, WHO estimates that as many as 20% may be due to UV exposure. Experts believe that each 1% sustained decrease in stratospheric ozone would result in an increase of 0.5% in the number of cataracts caused by solar UV. Direct viewing of the sun and other extremely bright objects can also seriously damage the very sensitive part of the retina called the yellow spot, fovea or macula leutea. When cells of the fovea are destroyed, people can no longer view fine detail. For those people it becomes impossible to read, sew, watch TV, recognize faces, drive a vehicle etc.

8.3.4 Immune System

UV also appears to alter immune response by changing the activity and distribution of the cells responsible for triggering these responses. A number of studies indicate that UV exposures at environmental levels suppress immune responses in both rodents and humans. In rodents, this immune suppression results in enhanced susceptibility to certain infectious diseases with skin involvement, and some systemic infections. Mechanisms associated with UV-induced immunosuppression and host defence that protect against infectious agents are similar in rodents and humans. It is therefore reasonable to assume that UV exposure may enhance the risk of infection and decrease the effectiveness of vaccines in humans. Additional research is necessary to substantiate this.

8.3.5 UV Index

The Global Solar UV Index was developed through the WHO. It provides an estimate of the maximum solar UV exposure at the Earth's surface. The intensity of UV reaches a maximum around mid-day (when there is no cloud cover) at solar noon.

It is generally presented as a forecast of the maximum amount of skin-damaging UV expected to reach the Earth's surface at solar noon. The values of the Index range from zero upward; the higher the Index number, the greater the likelihood of skin and eye damaging exposure to UV, and the less time it takes for damage to occur.

Close to the equator, summer-time values reach 20. During a European summer a value of 8 can be reached. We speak of:

- low UV exposure: Index 1...2
- moderate UV exposure: Index 3...4
- high UV exposure: Index 5...6
- very high UV exposure: Index 7...8
- extreme UV exposure: Index > 9

8.4 Spacesuits

Outer space is extremely hostile and without a spacesuit the following things would happen:

- you would become unconscious within 15 s because there is no O,
- blood and other body fluids start to boil and then freeze because there is no air pressure,
- tissues (skin, heart...) expand because of the boiling fluids,
- extreme temperature changes: sunlight 120^0C, shade -100^0 C.
- exposure to radiation and micrometeoroids.

8.4.1 The Extravehicular Mobility Unit

Some facts: Weight = 127 kg on Earth, Thickness = 0.48 cm, 13 layers, atmosphere = 0.29 atm of pure oxygen, Volume = 0.125 to 0.153 m^3, without astronaut cost = 12 million USD.

While early spacesuits were made entirely of soft fabrics, the EMU has a combination of soft and hard components to provide support, mobility and comfort. The suit itself has 13 layers of material, including an inner cooling garment (two layers), pressure garment (two layers), thermal micrometeroid garment (eight layers) and outer cover (one layer). The materials used include: Nylon tricot Spandex, Urethane-coated Nylon, Dacron, Neoprene-coated Nylon, Mylar, Gortex, Kevlar (material in bullet-proof vests), Nomex. All of the layers are sewn and cemented together to form the suit. In contrast to early spacesuits, which were individually tailored for each astronaut, the EMU has component pieces of varying sizes that can be put together to fit any given astronaut. The EMU consists of the following parts:

- Maximum Absorption Garment (MAG) - collects urine produced by the astronaut. Liquid Cooling and Ventilation Garment (LCVG) - removes excess body heat produced by the astronaut during spacewalks EMU.

- Electrical Harness (EEH) - provides connections for communications and bio-instruments.

- Communications Carrier Assembly (CCA) - contains microphones and earphones for communications.

- Lower Torso Assembly (LTA) - lower half of the EMU including pants, knee and ankle joints, boots and lower waist Hard Upper Torso (HUT) - hard fiberglass shell that supports several structures including the arms, torso, helmet, life-support backpack and control module Arms Gloves - outer and inner gloves Helmet.

- Extravehicular Visor Assembly (EVA) - protects the astronaut from bright sunlight

- In-suit Drink Bag (IDB) - provides drinking water for the astronaut during the spacewalk.

- Primary Life Support Subsystem (PLSS) - provides oxygen, power, carbon dioxide removal, cooling water, radio equipment and warning system.

- Secondary Oxygen Pack (SOP) - provides emergency oxygen supply.

- Display and Control Module (DCM) - displays and controls to run the PLSS

8.5 Radiation Shielding

Since the 1950s it is known that radiation in space poses a problem to human space travel. In 1952 Wernher von Braun and other space visionaries suggested using lunar soil to protect manned expedition from space radiation and meteors.

Low energy radiation can be stopped by a spacecraft wall. At higher energies the wall itself produces showers of secondary radiation and even more shielding is needed to absorb that. Using light weight materials like hydrogen, boron and lithium, nuclei of heavy elements in cosmic rays can be shattered by lightweight atoms without producing additional hazardous recoil products like neutrons. Thus, composites and other materials using low mass atoms might provide good shielding. At NASA's Langley Research Center simulated Mars soil will be tested for shielding.

The International Space Station (ISS) at 51.6^0 inclination and 220 mile of altitude is being constructed during a period of high solar activity with about 1000 hours of required extra vehicular activity (EVA). The Astronauts are exposed to trapped protons and electrons and galactic cosmic rays. Especially during transits through the South Atlantic Anomaly (SAA) during EVA astronauts may experience enhanced doses. Dose enhancements are also expected from solar particle events (SPE). There are two different types of suits for astronauts: EMU and Orlan.

The skin responses to radiation include erythema, epilation, desquamation. Different anatomical skin sites vary in sensitivity with decreased order of responsiveness as follows:

Table 8.5: Common shielding materials

lead	11.35 g/cm^3
aluminium	2.7
water	1.0
lithium hydride	0.82
liquid hydrogen	0.07

- anterior aspect of neck,
- anterior surfaces of extremities, chest, abdomen,
- face
- back, posterior surfaces,
- nape of neck,
- scalp, palms, soles.

Literature: For more details on the problem see e.g. Kiefer (2001) or the Space Studies Board of the National Reserach Council (2000) or Thomson (1999), Badhwar (1997). Radiation measurements on Russian spacecraft Mir are presented by MacKay et al. (1993).

The effectiveness of any shielding depends on the energy distribution of the incident radiation. Some examples for common shielding materials are listed in Table 8.5.

Radiation with energy less than 1 MeV/nm can not penetrate a space suit of 1 mm thickness. Al shielding reduces the low boundary to 40 MeV. When a high-energy ion strikes an atom in metal shielding it can produce secondary radiation and there are cases where a small amount of shielding is worse than none at all. Bremsstrahlung can be created (X-rays) by electrons as they interact with spacecraft shielding.

Chapter 9

The Ionosphere and Space Weather

9.1 General Properties

The ionosphere contains only a small fraction of the Earth's atmosphere (above 100 km less than 1 % of the mass of the atmosphere). However, this layer is extremely important for modern telecommunication systems since it influences the passage of radio waves.

Because of its name we can expect that the atoms are ionized there. Most of the ionosphere is electrically neutral. On the sunlit side of the Earth the shorter wavelengths of solar radiation (extreme UV and X rays) are energetic enough to produce ionization of the atoms. Therefore, this layer becomes an electrical conductor supporting electric currents and radio wave propagation.

Historically, it has been divided into regions with specific ionizations.

- lowest D region: between 50 and 90 km.
- E region: between 90 and 150 km,
- F region: contains the F1 and F2 layers.

The top of the ionosphere is at about 1000 km, however there exists no definite boundary between plasma in the ionosphere and the outer reaches of the Earth's magnetic field. In the E region the most important ions are O_2^+, NO^+, in the F region it is O^+. In the F2 layer (at about 400 km) the electron concentration reaches its highest values which is important for the telecommunication systems. At high latitudes there is another source of ionization of the ionosphere– the aurora. The light of the aurora is caused by high speed electrons and protons, coming from the magnetosphere spiralling down the Earth's magnetic field lines.

At low latitudes the largest electron densities are found in peaks on either side of the magnetic equator, which is called the equatorial anomaly. Normally one would expect that the peak concentration will occur at the equator because of the

Table 9.1: Variation of the ionosphere

Ionospheric parameter	Diurnal (Mid-Latitude)	Solar Cycle (daytime)
N_{max}	$1 \times 10^5 ... 1 \times 10^6 \, e^-/cm^3$	$4 \times 10^6 .... 2 \times 10^6 \, e^-/cm^3$
	Factor of 10	Factor of 5
Max. Usable Freq. MUF	12...46 MHz	21 ...42 MHz
	Factor of 3	Factor of 2
Total Electron Content TEC	$5...50 \times 10^{16} \, e^-/m^2$	$10...50 \times 10^{16} \, e^-/m^2$
	Factor of 10	Factor of 5

maximum in solar ionizing radiation. This peculiarity can be explained by the special geometry of the magnetic field and the presence of electric fields. The electric fields transport plasma and are caused by a polarizing effect of thermospheric winds.

The ionosphere varies because of two reasons:

- two varying sources of ionization (aurora, Sun)

- changes in the neutral part of the thermosphere, which responds to solar EUV radiation.

Thus the ionospheric variation mainly occurs at a 24 h period (daytime-nighttime) and over the 11 year cycle of solar activity. We observe considerable changes in the F-region maximum density (N_{max}) of the electrons which influences the plasma frequency that is proportional to it. On shorter time scales solar X-ray radiation changes dramatically during a solar flare eruption. This effect increases the D and E ionization. During a geomagnetic storm the auroral source of ionization becomes more intense. In extreme cases aurorae can be seen at moderate latitudes (Italy, Mexico). Another source of variability in the ionosphere comes from the interaction of charged particles with the neutral atmosphere in the thermosphere. Thermospheric winds can push the ionosphere along the inclined magnetic field line to a different altitude. Moreover the composition of the thermosphere affects the rate that ions and electrons recombine. During a geomagnetic storm energy input at high latitudes produces waves and changes in the thermospheric winds and composition. The electron concentration can increase (positive phases) and decrease (negative phases). The ionospheric variability is given in Table 9.1.

HF communication depends on radio waves that are reflected in the ionosphere. This is characterized by the maximum usable frequency (MUF) and the lowest usable frequency (LUF). The MUF depends on the peak electron density in the F region and the angle of incidence of the emitted radio wave. As we have seen, this changes during the day, over the solar cycle and during geomagnetic disturbances. The LUF is controlled by the amount of absorption of the radiowave in the lower D and E layers. This is severely affected by solar flares. All single frequency GPS

9.1. GENERAL PROPERTIES

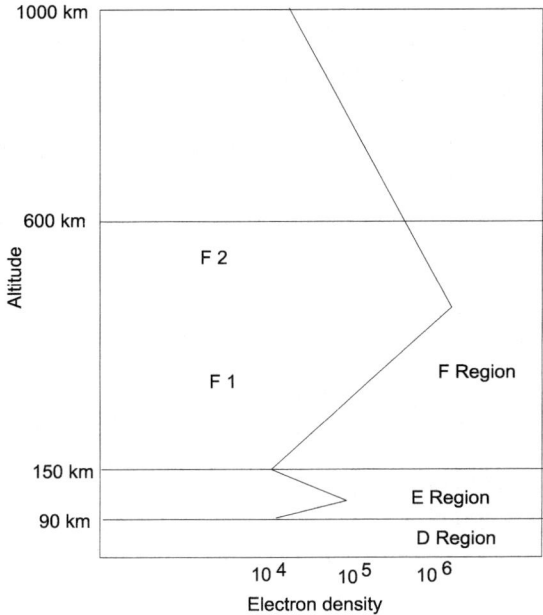

Figure 9.1: The Earth's ionosphere

receivers must correct the delay of the GPS signal as it propagates through the ionosphere to the GPS satellite (at 22 000km altitude).

The ionosphere may become highly turbulent, mainly in the high latitude and low latitude F region and at special times (often after sunset). In this context turbulence is defined as small scaled structures (scale length cm to m) which are irregular and embedded in the large scale ambient ionosphere (tens of kilometers). In the equatorial region plasma irregularities are generated just after sunset and may last for several hours. At high latitudes these irregularities may be generated during day and night. Both effects occur most frequently during the solar cycle maximum. Radio signals become disrupted by these perturbations and the effect is known as *ionospheric scintillation*. The bigger the amplitude of the scintillated signal the greater the impact on communication and navigation systems.

9.1.1 Aurora

Energetic particles from the Sun are carried out into space along with the ever present hot solar wind. This wind sweeps toward Earth at super sonic speeds ranging from 300 to 1000 km/s. It distorts the Earth's magnetic field which forms out a comet shaped magnetosphere.

The magnetosphere protects us from energetic particles and most of them are deflected around the Earth. The Earth's *Van Allen Belts* consist of highly energetic ionized particles trapped in the Earth's geomagnetic field. On the sunward side of

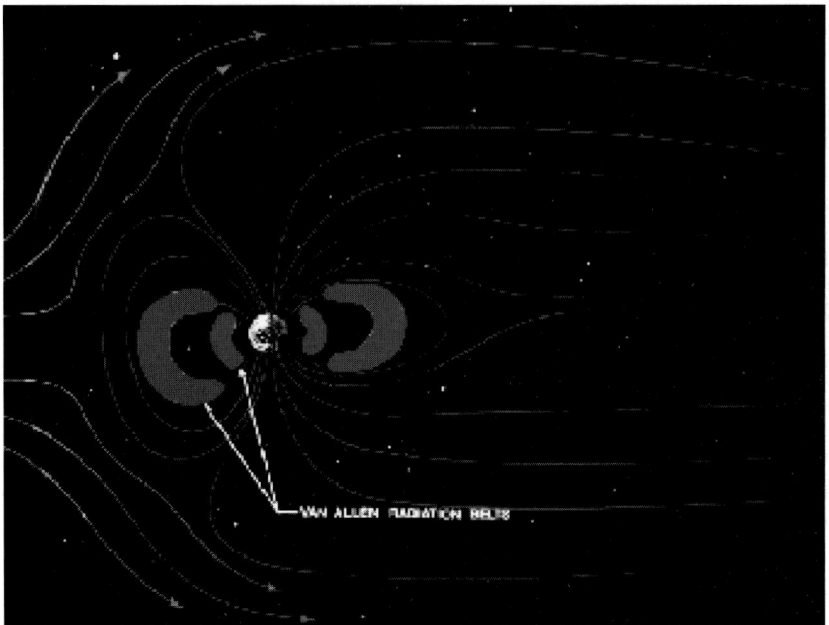

Figure 9.2: Van Allen Belts

the Earth, the geomagnetic field is compressed by the solar wind, on the opposite side of the Earth, the geomagnetic field extends. Thus the field forms an elongated cavity which is also known as the Chapman-Ferraro Cavity around the Earth. Within this cavity are the Van Allen Radiation Belts. The radiation belts are composed of electrons (keV) and protons (MeV).

As it is seen in Figure 9.2 there are two belts of particle concentration: a) small inner belt between 1 and 2 Earth radii where protons of energy 50 MeV and electrons with energies $>$ 30 MeV reside and b) outer larger belt from 3 to 4 Earth radii where less energetic protons and electrons are concentrated. The inner belt is relatively stable, the outer belt varies in its number of particles by as much as a factor of 100.

Charged particles trapped in the belts spiral along the field lines while bouncing between the northern and southern mirror points. Particles in the inner belt may interact with the upper atmosphere causing the auroral oval which is an annulus centered over the magnetic poles and around 3000 km in diameter during quiet times. The location of the auroral oval is usually found between 60 and 70 degrees of magnetic latitude (north and south).

There are many shapes and features of aurorae. They generally start at 100 km above the surface and extend upward along the magnetic field for hundreds of km. Auroral arcs can nearly stand still and then suddenly move (dancing, turning). After midnight one often sees a patchy appearance of aurorae, and the patches blink on and off every 10 s or so. Most of aurorae are greenish yellow and

9.1. GENERAL PROPERTIES

sometimes the tall rays turn red at their top and along their lower edge. On rare occasions sunlight hits on the top creating a faint blue color.

The different colors depend on the specific atmospheric gas, its electrical state and on the energy of the particle that hits the atmospheric gas. Atomic oxygen is responsible for the two main colors of green (557.7 nm, at a height below 400 km) and red (630.0 nm, about 400 km or higher). Excited nitrogen also emits light (600-700 nm; below 200 km). Auroral displays are intensified if the interplanetary magnetic field is in the opposite direction to the Earth's magnetic field. The geomagnetic storms produce brightness changes and motion in the aurorae and these are called auroral substorms. Recent models of aurorae explain the phenomenon by a process of release of energy from the magnetotail, called magnetic reconnection. Regions of opposite magnetic fields come together and the magnetic field lines can break and reconnect in new combinations. The point of reconnection in the magnetotail lies usually at 100 Earth radii. When the solar wind adds sufficient magnetic energy to the magnetosphere, the field lines there overstretch and a new reconnection takes place at 15 Earth radii, the field collapses and electrons are injected into the atmosphere.

Reconnection stores large amounts of energy in the Earth's magnetic field until it is released explosively. The cycle of energy storage and release is called substrom. Multiple substorms lead to magnetic storms and acceleration of particles to very high energies. These particles damage satellites.

The geomagnetic field is measured by magnetometers and the data are often given as 3-hourly indices that yield a quantitative measure of the level of geomagnetic activity. The K-index is given from 0 to 9 and depends on the observing station. The globally averaged K_p index is a measure for the global auroral activity.

When geomagnetic activity is low, the aurora typically is located at about 67 degrees magnetic latitude, in the hours around midnight. As activity increases, the region of aurora expands towards the equator. When geomagnetic activity is very high, the aurora may be seen at mid and low latitude locations (see Table 9.2) around the earth that would otherwise rarely experience the polar lights.

In Table 9.3 auroral boundaries are given as a function of the Kp index.

The magnetic activity produced by enhanced ionospheric currents flowing below and within the auroral oval is measured by the Auroral Electrojet Index AE. The definition of this index is as follows: at a certain time the total range of deviation from quiet day values of the horizontal magnetic field (h) around the auroral oval. Defined and developed by Davis and Sugiura, AE has been usefully employed both qualitatively and quantitatively as a correlative index in studies of substorm morphology, the behavior of communication satellites, radio propagation, radio scintillation, and the coupling between the interplanetary magnetic field and the Earth's magnetosphere. For these varied topics, AE possesses advantages over other geomagnetic indices or at least shares their advantageous properties.

Table 9.2: Corrected magnetic latitudes of some cities

Atlanta	44.5	Athens	31.3	Adelaide	45.9
Boston	51.7	Berlin	48.3	Buenos Aires	23.3
Chicago	52.2	Copenhagen	51.9	Capetown	41.5
Dallas	42.7	Edinburgh	53.0	Christchurch	49.9
Denver	48.3	London	47.5	Comodoro Rivadavia	32.1
Great Falls, MT	54.9	Madrid	33.3	Concepcion, Chile	23.2
Havana	34.1	Moscow	51.8	Dunedin	53.0
Los Angeles	39.8	Paris	44.2	Durban	38.8
Mexico City	29.1	Perm	53.8	East London	41.1
Minneapolis	55.1	Prague	45.5	Hobart	53.6
New York	50.6	Rome	35.5	Melbourne	48.4
Quebec City	56.2	St. Petersburg	56.1	Perth	43.9
San Francisco	42.5	Warsaw	46.7	Punta Arenas, Chile	38.6
Seattle	52.7	Beijing	34.1	Sydney	43.5
St. Louis	49.2	Irkutsk	47.0	Toronto	53.9
Seoul	31.0	Washington, DC	49.1	Tokyo	29.0
Winnipeg	59.5	Vladivostok	36.5	Vienna	43.0

Table 9.3: Extension of the auroral zone. The first values given is the magnetic latitude (Lat), the second the Kp index.

Lat	Kp	Lat	Kp	Lat	Kp	Lat	Kp	Lat	Kp
66.5	0	64.5	1	62.4	2	60.4	3	58.3	4
56.3	5	54.2	6	42.2	7	50.1	8	48.1	9

9.1.2 Geomagnetic indices

Daily regular magnetic field variation arise from current systems caused by regular solar radiation changes. Other irregular current systems produce magnetic field changes caused by

1. the interaction of the solar wind with the magnetosphere,

2. by the magnetosphere itself,

3. by the interactions between the magnetosphere and ionosphere,

4. and by the ionosphere itself.

Therefore, magnetic activity indices were designed to describe variation in the geomagnetic field caused by these irregular current systems.

Let us give a brief description of other geomagnetic indices which are interesting for the solar-terrestrial relations.

9.1. GENERAL PROPERTIES

Table 9.4: Transformation between the Kp and the Ap index

Kp	0o	0+	1-	1o	1+	2-	2o	2+	3-	3o	3+	4-	4o	4+
Ap	0	2	3	4	5	6	7	9	12	15	18	22	27	32
Kp	5-	5o	5+	6-	6o	6+	7-	7o	7+	8-	8o	8+	9-	9o
Ap	39	48	56	67	80	94	111	132	154	179	207	236	300	400

Table 9.5: Transformation between the Ap and the Cp index

Cp	0.0	0.1	0.2	0.3	0.4	0.5	0.6	0.7	0.8	0.9	1.0
Ap	2	4	5	6	8	9	11	12	14	16	19
Cp	1.1	1.2	1.3	1.4	1.5	1.6	1.7	1.8	1.9	2.0	
Ap	22	26	31	37	44	52	63	80	110	160	

DST Index

DST stands for Disturbance Storm Time. The DST is an index of magnetic activity derived from a network of near-equatorial geomagnetic observatories that measures the intensity of the globally symmetrical equatorial electrojet (the "ring current"). Thus DST monitors the variations of the globally symmetrical ring current, which encircles the Earth close to the magnetic equator in the Van Allen (or radiation) belt of the magnetosphere. During large magnetic storms the signature of the ring current can be seen in ground magnetic field recordings worldwide as so-called main phase depression. The ring current energization which results in typical depression of 100 nT is related to magnetic reconnection processes at the neutral sheet.

Kp and Ap Index

The *K-Index* was first introduced by J. Bartels in 1938. It is a quasi-logarithmic local index of the 3-hourly range in magnetic activity relative to an assumed quiet-day curve for a single geomagnetic observatory site. The values consist of a single-digit 0...9 for each 3-hour interval of the universal time day (UT).

The planetary 3-hour-range index *Kp* is the mean standardized K-index from 13 geomagnetic observatories between 44 degrees and 60 degrees northern or southern geomagnetic latitude. The scale is 0...9 expressed in thirds of a unit, e.g. 5- is 4 2/3, 5 is 5 and 5+ is 5 1/3. This planetary index is designed to measure solar particle radiation by its magnetic effects. The 3-hourly Ap (equivalent range) index is derived from the Kp index (see Table 9.4). This table is made in such a way that at a station at about magnetic latitude 50 degrees, Ap may be regarded as the range of the most disturbed of the three field components, expressed in the unit of 2 g. A daily index *Ap* is obtained by averaging the eight values of Ap for each day. The *Cp* index, the daily planetary character figure, is defined on the basis of Ap according to Table 9.5

Another index devised to express geomagnetic activity on the basis of the Cp

Table 9.6: Transformation between the Cp and the C9 index

Cp	0.0-0.1	0.2-0.3	0.4-0.5	0.6-0.7	0.8-0.9
C9	0	1	2	3	4
Cp	1.0-1.1	1.2-1.4	1.5-1.8	1.9	2.0-2.5
C9	5	6	7	8	9

index is the $C9$ index which has the range between 0 and 9. The conversion table from the Cp index to the C9 index is given by 9.6

AE and other indices

These indices describe the disturbance level recorded by auroral zone magnetometers.

In order to determine these indices, horizontal magnetic component recordings from a set of globe-encircling stations are plotted to the same time and amplitude scales relative to their quiet-time levels. They are then graphically superposed. The upper and lower envelopes of this superposition define the AU (amplitude upper), the AL (amplitude lower) indices and the difference between the two envelopes determine the AE (Auroral Electrojet) index, i.e., AE = AU - AL. AO is defined as the average value of AU and AL.

Summary of geomagnetic indices

A summary of the indices as well as a few other indices can be found in Table 9.7

9.1.3 Solar Indices

10.7 cm Radio Flux

The sun emits radio energy with slowly varying intensity. This radio flux, which originates from atmospheric layers high in the sun's chromosphere and low in its corona, changes gradually from day to day in response to the number of spot groups on the disk. Solar flux from the entire solar disk at a frequency of 2800 MHz has been recorded routinely by a radio telescope near Ottawa since February 1947. The observed values have to be adjusted for the changing Sun-Earth distance and for uncertainties in antenna gain (absolute values). Fluxes are given in units of 10^{-22} $Js^{-1}m^{-2}Hz^{-1}$.

Sunspot Numbers

The sunspot number index is also often called Wolf number in reference to the Swiss astronomer J. R. Wolf who introduced this index in 1848; details about how to obtain that number can be found in the chapter about sunspots and the solar cycle.

Table 9.7: Summary of geomagnetic indices

aa	3-hour range index, derived from two antipodal stations
AE, AU, AL	1- , 2.5-minute, or hourly auroral electrojet indices
	am, an, as 3-hour range (mondial, northern, southern) indices
Ap	3-hour range planetary index derived from Kp
C, Ci, C9	Daily local (C) or international (Ci) magnetic character; C9 was first derived from Ci, then from Cp
Cp	Daily magnetic character derived from Kp
Dst	Hourly index mainly related to the ring current
K	3-hour local quasi-logarithmic index
Km	3-hour mean index derived from an average of K indices (not to be confused with the Km of the next item)
Km, Kn, Ks	3-hour quasi-logarithmic (mondial, northern, southern) indices derived from am, an, as
Kp, Ks	3-hour quasi-logarithmic planetary index and the intermediate standardized indices from which Kp is derived (not to be confused with the Ks of the preceding item)
Kw, Kr	3-hour quasi-logarithmic worldwide index and the intermediate from which Kw is derived
Q	Quarter hourly index
R	1-hour range index
RX, RY, RZ	Daily ranges in the field components
sn, ss	3-hour indices associated with an and as
U, u	Daily and monthly indices mainly related to the ring current
W	Monthly wave radiation index

9.1.4 Navigation Systems

Modern travel requires exact latitude, longitude and altitude information in real time. Therefore terrestrial based radio wave systems such as the Loran-C and the Omega-system were developed. They use large transmitter antennas to send low-frequency (LF) and very-low-frequency (VLF) radio signals along the ground and off the reflective layer provided by the ionosphere. Thus, vast distances over land and sea can be reached. More recently, space-based systems have become the tools for navigation, among others the GPS system (Global Positioning System). The advantage of space-based systems is that the satellites can easily cover the globe. A user can obtain an accurate three dimensional position (his location and altitude) as soon as at least four satellites are in view.

However both navigational systems, space-systems as well as systems on the surface suffer from the transmission through the ionosphere. The Omega system requires it, the Loran system tries to avoid it and the GPS system depends on radio signals that pass through it. Flares produce X rays and we have already discussed the influence of this shortwave radiation on the D and E region in the

Table 9.8: Navigation systems

System	Frequency	
Omega	VLF, kHz	about 10^4 Hz
Loran-C	LHF	about 10^5 Hz
GPS	UHF, GHz	about 10^9 Hz

ionosphere. Navigation with Loran-C and Omega systems thus is influenced by these events and during the maximum phase of the solar cycle daylight users of Loran-C and Omega systems have more difficulties. The GPS system is not influenced by this perturbation. The GPS operations are affected by the total electron content of the ionosphere along the path to the satellite and are thus influenced by geomagnetic storms. Whereas solar X-rays impact only the sunlit hemisphere of Earth, geomagnetic storms are ubiquitous. The ionospheric response to the storms also depends on the latitude. The conditions nearer to the equator or nearer to the poles vary for the user. It must also be stressed that a quiet undisturbed geomagnetic field does not necessarily dictate an undisturbed equatorial ionosphere. The influence of TEC variation (Total Electron Content) on GPS receivers is smaller for dual band receivers which actually measure the effect of the ionosphere on the GPS signals and correct the resulting positions for these. Unpredictable density enhancements can occur in the evening hours and cause scintillations which affect both dual- and single-frequency GPS receivers.

We summarize the effect of the space environment on the navigation systems:

- Loran-C: Phase and amplitude shifts due to skywave interference at the limits of coverage area.

- GPS: Carrier loss-of-lock due to ionospheric density fluctuations with solar or geomagnetic activity.

- Omega: Phase anomalies due to varying ionospheric reflection height; caused by solar or geomagnetic activity.

9.1.5 Radio Communication

The ionosphere affects the propagation of radio signals in different ways depending on their frequencies. Frequencies below 30 MHz are reflected in the ionosphere; this allows radio communication to distances of many thousands of kilometers. Radio signals at frequencies above 30 MHz penetrate the ionosphere and are useful for ground-to-space communications. Frequencies between 2 and 30 MHz are affected by increased absorption, higher frequencies by different reflection properties in the ionosphere. TV and FM radio stations (on VHF) are affected little by solar activity. HF ground to air, ship to shore, amateur radio etc. are affected strongly. Also the Faraday rotation of the plane of polarization has to be taken into account (for satellite which employ linear polarization up to 1 GHz).

9.1. GENERAL PROPERTIES

During a solar flare event a sudden increase of X-ray emission causes a large increase in ionization in the lower regions of the ionosphere on the sunlit side of the Earth. Very often one observes a sudden ionospheric disturbance (SID). This affects very low frequencies (OMEGA) as a sudden phase anomaly (SPA) or a sudden enhancement of the signal (SES). At HF and sometimes also at VHF an SID may appear as a short wave fade (SWF). Depending on the magnitude of the solar flare such a disturbance may last from minutes to hours. At VHF the radio noise created by solar flares interferes with the signal. The occurrence of solar flare is modulated by the solar activity.

Flares may also emit energetic particles. The PCA (polar cap absorption) is caused by high energetic particles that ionize the polar ionosphere. A PCA may last from days to weeks depending on the size of the flare and the interaction of the high energetic particles emitted by the flare and the Earth's magnetosphere. During these events polar HF communication becomes impossible. A coronal mass ejection may be a consequence of a large solar flare or a disappearing filament and is an ejection of a large plasma cloud into the interplanetary space. Such a coronal mass ejection (CME) travels through the solar wind and may also reach the Earth. This results in a global disturbance of the Earth's magnetic field and is known as a geomagnetic storm. High speed solar wind streams originating in coronal holes on the Sun's corona hits the Earth's magnetosphere and also causes ionospheric disturbances.

9.1.6 Geomagnetically Induced Currents

Ground effects of space weather are generally known as GIC (geomagnetically induced currents). These currents are driven by the geoelectric field associated with a magnetic disturbance in electric power transmission grids, pipelines, communication cables and railway equipment. GIC are dc currents. They may cause several effects because they increase existing current and this may cause saturation:

- Increase of harmonics,
- unnecessary relay trippings,
- increase in reactive power loss,
- voltage drops,
- permanent damage to transformers,
- black out of the whole system.

When flowing from the pipeline into the soil, GIC may increase corrosion of the pipeline, and the voltages associated with GIC disturb the cathodic protection system and standard control surveys of the pipeline.

On March 13, 1989, the most famous GIC failure occurred in the Canadian Hydro-Quebec system during a great magnetic storm on March 13, 1989. The system suffered from a nine-hour black-out.

A theoretical calculation of GIC in a given network (power grid, pipeline etc.) can be divided into two steps:

- Calculate the geoelectric field created primarily by ionospheric-magnetospheric currents and affected secondarily by the earth's conductivity distribution. This is also called the geophysical step.

- Calculate the currents produced by the geoelectric field in the circuit system constituted by the network and its earthings.

The first step is generally more difficult, partly because the space and geophysical input parameters are not well known.

The effects of geomagnetic disturbances on electrical systems at the earth's surface were studied e.g. by Boteler et al. (1998) or Lehtinen and Pirjola (1985). A prediction of Geomagnetically Induced Currents in Power Transmission Systems was given by Pirjola et al. (2000).

9.1.7 Systems Affected by Solar or Geomagnetic Activity

In this paragraph we give a summary of the influence of solar and geomagnetic activity (driven by solar events) on various systems.

- HF Communications
    - Increased absorption
    - Depressed MUF
    - Increases LUF
    - Increases fading and flutter

- Surveillance Systems
    - Radar energy scatter (auroral interference)
    - Range errors
    - Elevation angle errors
    - Azimuth angle errors

- Satellite Systems
    - Faraday rotation
    - Scintillation
    - Loss of phase lock
    - Radio Frequency Interferences (RFI)

- Navigation Systems
    - Position errors

9.2. SATELLITES

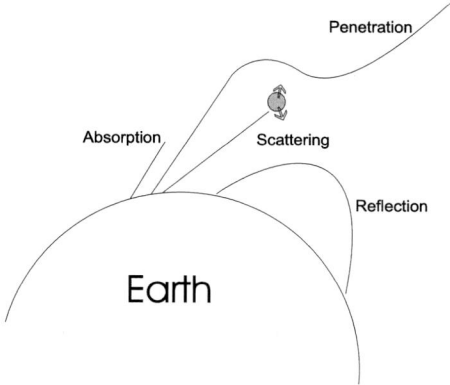

Figure 9.3: Radio signal propagation in the ionosphere.

9.2 Satellites

9.2.1 Solar Panels

Solar panels are devices that convert light into electricity. Some scientists call them photovoltaics which means, basically, "light-electricity" since the solar light is converted into electric energy by them.

A solar panel is a collection of solar cells. Lots of small solar cells spread over a large area can work together to provide enough power for satellites or space stations. The more light that hits a cell, the more electricity it produces, so spacecrafts are usually equipped with solar panels that can always be pointed at the Sun even as the rest of the body of the spacecraft moves around.

The most efficient solar panels are the DS1 solar panels which convert about 22 % of the available energy into electrical power whereas most solar panels on people's houses convert only about 14 %. It is also important to note that solar panels lose about 1-2 % of their effectiveness per year. This means after a five year mission, the solar panels will still be making more than 90 % of what they made at the beginning of the mission. Of course this also depends on their distance to the Sun.

There are two major dangers to solar panels in space besides regular wear-and-tear:

- Solar flares that can damage the electronics inside the panels.

- Micrometeorites, which are tiny, gravel-sized bits of rock and other space junk floating in space can scratch or crack solar panels.

Some protection can be made by the use of a thick layer of glass. Of course, if a satellite's mission path takes it away from the Sun (further out into the solar system) solar panels will become less and less efficient.

Another kind of protection to the above mentioned damaging effects can be made by the use of a solar concentrator. This uses Fresnel lenses which collect a large area of sunlight and direct it towards a specific spot by bending the rays of light and focussing them- the same principle when people use a magnifying lens to focus the Sunlight on a piece of paper which starts a small fire.

Fresnel lenses have been invented in 1822 by Jean Fresnel. Theaters use them for spotlights. They are shaped like a dart board with concentric rings around a lens that is a magnifying glass. Solar concentrators put one of these lenses on top of every solar cell. The solar cells can then be spaced farther apart since the light is focused on each cell. Fewer cells need to be placed and the panels cost less to construct. Thick glass or plastic cover over the solar panel are used to protect them from micrometeorites.

DS1's photovoltaics are made out of gallium arsenide (GaAs). GaAs is made into a cylinder that is then sliced into cells. These solar cells are then connected to the rest of the power network. Solar concentrators, made of clear plastic, are placed above them to focus the Sun's rays.

As a summary we give some literature, further references can be found therein. Markvart et al. (1982) studied the photon and electron degradation of boron-doped FZ silicon solar cells. Radiation-resistant silicon solar cell were investigated by Markvart et al. (1987). Defect interactions in silicon solar cells were analyzed by Markvart et al. (1989). A study of radiation-induced defects in silicon solar cells showing improved radiation resistance was made by Peters et al. (1992). General information about solar cells can be found in Tada et al. (1982).

A review on radiation damage in solar cells was given by Markvart (1990).

9.2.2 Power Sources for Spacecraft

Every power source available for a satellite or other spacecrafts has different strengths and weaknesses. By combining different power sources one can reach an optimum in power generation.

- Batteries: a reliable, well understood technology. However, power demands for satellites tend to be very high and a battery that would be strong enough to power a satellite for the length of a mission would be larger than the satellite itself. Thus, batteries are used as a temporary storage for power from another source.

  A battery can convert chemical energy to electricity by putting certain chemicals in contact with each other in a specific way. Electrons will travel from one kind of chemical to another creating an electric current.

  Batteries come in several styles; everyone is probably most familiar with single-use alkaline batteries. NASA spacecraft usually use rechargeable nickel-cadmium or nickel-hydride batteries like those found in laptop computers or cellular phones (DS1 uses nickel-hydrogen batteries). Engineers think of batteries as a place to store electricity in a chemical form.

  Batteries tend to expend their charge fairly quickly. DS1 can last from half an hour to three hours running purely on battery power before the batteries need to be recharged from the solar panels. These batteries are recharged thousands of times during the life of the spacecraft.

9.2. SATELLITES

Table 9.9: Fuels for RTG's

Element	Half life (years)	Watts/g (thermal)	Watt (thermal)
^{210}Po	0.378	141	570
^{238}Pu	86.8	0.55	3000
^{144}Cs	0.781	25	15
^{190}Sr	28.0	0.93	250
^{242}Cm	0.445	120	495

- Solar panels: they provide abundant power for nearly all a satellite's needs and are safe and clean to launch. However:

  - solar panels are large and fragile constructions that are vulnerable to damage from external forces or even mechanical failures;
  - they are rather expensive to build and put into space;
  - they always need to be pointed at the Sun (think about what happens if they are blocked by planets or other objects);
  - the farther the satellite gets from the Sun, the less effective solar panels work. As a rule of thumb we can state that solar powered missions cannot travel further than the orbit of Mars.

- Radioisotope thermoelectric generators: They are also reliable but tend to be expensive to build and of course there is a risk that radioactive material is set into the environment during a launch failure.

  A radioisotope thermoelectric generator, or RTG, uses the fact that radioactive materials (such as plutonium) generate heat as they decay. The heat is converted into electricity by an array of thermocouples which then power the spacecraft.

  A thermocouple is a device which converts thermal energy directly into electrical energy. Basically, it is made of two kinds of metal that can both conduct electricity. They are connected to each other in a closed loop. If the two metals are at different temperatures, an electric potential will exist between them. When an electric potential occurs, electrons will start to flow, making electric current.

  Another process which belongs to this group of energy generation is nuclear fission where unstable radioactive materials are split into smaller parts. Very large amounts of heat are generated but the whole process is more complex and not as reliable as using the heat produced by radioactive decay. An RTG is steadier.

  Plutonium is a very toxic heavy metal. If it is powdered and inhaled, it is a cancer causing agent. It is sealed inside a hard, radiation proof shell. The shell is designed to survive all conceivable accidents, so even in the unlikely event of a launch failure, none of the radioactive particles should escape.

- Fuel cells: they are similar like batteries but they have a longer lifespan and can be refuelled. They are already in use in the Space Shuttle. However they run hot (400-800^0 C) and the waste heat is often hard to manage.

  When atoms of the two gases oxygen and hydrogen are put next to another, they spontaneously combine to form water. This results in the release of a lot of energy. In a fuel cell the H and O are separated by a membrane. The refuelling means just to provide more H and O and the waste is pure water. With an external source such as a solar panel, one can split the waste water back into its component parts and use it again as fuel. Fuel cells were first used by the Apollo missions since they last longer than traditional batteries and didn't have expensive radioactive parts.

It is extremely important to control the heat on and around a space ship. The operating temperature is usually given between two numbers like -10^0C to 60^0C. The parts of the spacecraft have been tested and will work if the temperature in the spacecraft is between these two numbers. Why does a spacecraft have an operating temperature? For example the rocket thruster can use hydrazine as rocket fuel. Therefore the tanks, plumbing and pumps must be kept at a certain temperature: Hydrazine freezes at 2^0C and boils at 113^0C. Most electronic components will work only within a narrow range of temperatures, usually -50^0C to $+150^0$C, components will stop working and make the spacecraft useless if the spacecraft temperatures become too extreme.

Heat tends to expand material parts and the opposite happens when a part is cooled. This problem occurs when one part of the spacecraft is pointed at the Sun and the other one is pointed at empty space. The Sun then heats up only one part and this uneven heating causes the spacecraft to be warped or even break or instruments can be distorted. Another source of heating is caused by electronic components. Heat also makes the electrical system less efficient. Electricity is caused by the flow of electrons. One electron will enter the electron cloud of an atom and another will leave and the continuing flow of electrons through the conductor is the electrical current. When the material is heated up the electron clouds will be not at the same place and sometimes an atom's nucleus will get in the way of the electrons; moreover, if an atom is heated up, it can hold more electrons for a longer period of time before sending extra electrons away and continuing the electrical flow. Heat sources can be external (from outside the spacecraft) or internal (from inside the spacecraft). External heat sources include:

- the Sun,
- reflected sunlight from planets and moons,
- heating by friction when travelling through an atmosphere or gas clouds,
- released heat from planets.

Internal heat is generated by the craft's propulsion or electrical system.

9.2.3 Satellite Crashes

Here we give some example of satellite crashes. The Compton Gamma Ray Observatory which had the size of a bus (15 t) has crashed safely back to Earth in June, 2000. It was brought down in a remote part of the Pacific Ocean after being sent orders to destroy itself by the US space agency NASA.

The satellite was taken out of commission because of fears it could crash to Earth in an uncontrolled re-entry over a populated area. It had problems with the gyroscopes which started half a year earlier.

One of the 70 Iridium telecommunication satellites, that were launched on Sep. 8, 1998 never worked properly and reentered Earth's atmosphere in Nov. 2000. Currently about 8000 artificial objects orbiting the Earth are tracked at the United States Space command. These objects comprise satellites, space debris and rocket bodies.

In 1978 a Soviet nuclear-powered satellite crashed into a lake in northeast Canada. There radioactive contamination occurred.

The Russian Mars-96 probe was the most ambitious Russian interplanetary expedition ever made. The launch was in November 1996. The spacecraft was so big that to get it into a parking orbit 160 km above the Earth required a huge three- stage Proton rocket as well as a kick from a fourth stage thruster that would remain attached to the vehicle. The fourth stage was then supposed to fire again an hour later to push the probe on its way to Mars. Even that would be insufficient to achieve the necessary escape velocity from the Earth's gravity. The probe would then separate from the expended fourth stage and fire its own small engine to complete the escape manoeuvre.

After the first burn of the fourth stage, Mars-96 reached its parking orbit 20 minutes after blastoff and as it crossed China, within direct radio contact of a Russian tracking site at Ussuriysk, near Vladivostok. It then crossed the Pacific from northeast to southwest, passing within 600 kilometers of the watching eyes of the American military space tracking site on Kwajalein Island in the mid-Pacific.

The second fourth-stage burn was supposed to occur as the spacecraft flew northeast over the coast of Uruguay on the eastern side of south America. The fourth stage failed and the rocket and the Mars-96 probe remained in the parking orbit.

At the time, nobody knew it had failed. The probe's autopilot never noticed the absence of the burn because a minute after the scheduled completion time, crossing the West African coast over the Côte d'Ivoire, it separated from the fourth stage and turned on its own small engine, as originally planned. This sent it into an elliptical orbit and towards its fiery demise.

None of this was yet known to the Russian trackers who had no contact with the spacecraft. Mars-96 finished its burn, and unfolded its solar arrays. At the main Russian space tracking facility at Yevpatoriya in the Crimea, the signal was received and for a few joyful moments the engineers thought that nothing was wrong.

But as the probe passed Yevpatoriya overhead, it became evident that the probe was trapped in an elliptical Earth orbit, with no possibility of an escape towards Mars. The mission was over. Only then began the desperate attempt to crash the spacecraft safely. But within minutes it had passed beyond the range of the Crimean station, and no other Russian site ever heard from it again.

The probe contained 200 grams of plutonium 238 and crashed into the pacific west of Easter island. People reported on an object that was brighter than Sirius with a luminous trail five degrees in length and fragmented.

9.2.4 Electron Damage to Satellites

During an explosive Solar Particle Event (SPE) satellites could suffer damage. These events are usually associated with solar flares and coronal mass ejections. Protons and electrons are emitted at high velocities which can cause problems in orbiting satellites. In January 1994 three geostationary satellites suffered failures of their momentum wheel control circuity. One of these satellites never fully recovered. During that period however, no SPE was observed. One explanation for this failure is done by assuming a long duration of high energy electron fluxes that occur during times of high speed solar wind streams. It is important to note that these occur during times of sunspot minimum. Thus not only the electron intensity but the total integrated electron flux is important. The USAF uses empirically defined values to issue warnings for satellite operators. Damaging conditions are assumed when the daily electron flux (which is given by the number of high energy electrons (> 2MeV) per cm^2 per sterad per day meets either of the following conditions:

- greater than 3×10^8 per day for 3 consecutive days; or
- greater than 10^9 for a single day.

Such conditions often occur about 2 days after the onset of a large geomagnetic storm.

How does damage to satellites occur? The important phenomenon here is Deep Dielectric Charging. High energy electrons penetrate the spacecraft's outer surface; they penetrate the dielectric materials such as circuit boards and the insulation in coaxial cables. This gives rise to intense electric fields; as soon as they exceed the breakdown potential of the material they produce sudden discharges (similar to a stroke). This discharge damages the system: components may start to burn, semiconductors may be destroyed. These dielectric charging can be avoided by a special construction of the relevant parts however this leads to additional weight and complexity of the system.

9.2.5 Single Event Upsets

Single-event upsets (SEUs) are random errors in semiconductor memory that occur at a much higher rate in space than on the ground. They are non-destructive, but can cause a loss of data if left uncorrected. SEUs are often associated with heavy ions from the galactic cosmic radiation.

What is the cause of SEUs? Energetic charged particles pass through sensitive regions of a chip. Depending on their energy and angle of impact, individual

9.2. SATELLITES

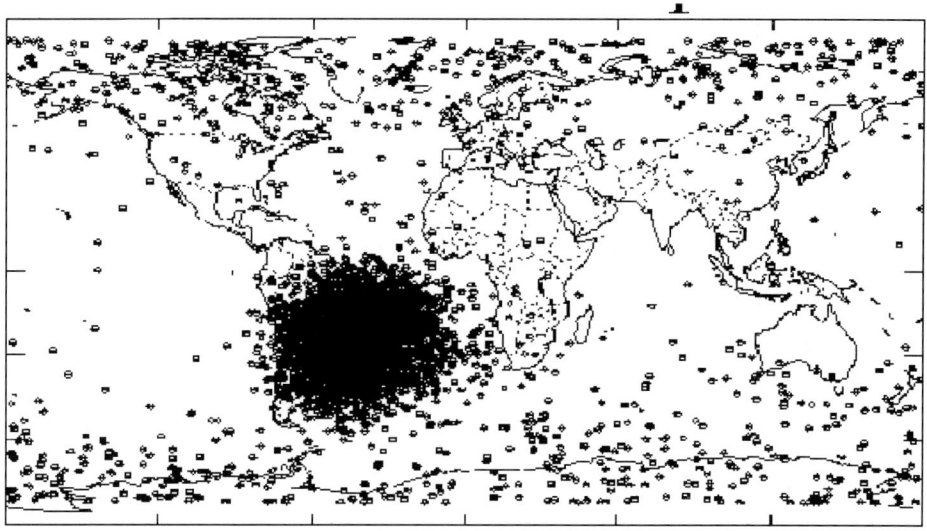

Figure 9.4: Single event upsets; spatial distribution of errors from the UoSAT-3 spacecraft in polar orbit; please note the South Atlantic Anomaly

particles can cause a large current impulse sufficient to change the state of a bistable circuit element.

Heavy Ion SEUs occur directly when a heavy ion passes through a semiconductor memory element. The standard models take into account the size, shape, and charge sensitivity of the memory element and the energy, angle, and impact parameter of the incident particle.

For satellites around the Earth, the offset and tilt of the geomagnetic axis with respect to the Earth's rotation axis produces a corresponding miss-alignment of the radiation belts. The result is the South Atlantic Anomaly. The spatial distribution of errors show a maximum in the Atlantic ocean east of the southern part of South America. There occurs also a significant number of errors at high latitudes due to cosmic rays (see Fig. 9.4). These data are from UoSAT-2 which measured from September 1988 to May 1992; UoSAT-2 monitored almost 9000 Single Event Upsets (SEU), and the majority of these (75%) occurred in the South Atlantic Anomaly (SAA) region.

Single event upsets pose also problems to space missions: As a result of volcanic action on Io, the innermost of the large Galilean moons of Jupiter, particles (actually heavy ions) of sulphur and oxygen are present in the space surrounding the planet. These particles form a part of the Jovian magnetosphere. Although the origin of these particles is the moon Io, the volcanoes provide enough velocity for them to escape from the gravitational field of the moon and to become elements of the magnetosphere around Jupiter.

The heavy ions diffuse both inward and outward from the planet. Many of the particles diffuse outward to 20 to 50 times the radius of Jupiter (R_J, measured from the planet's center), where they are accelerated by an interaction with the massive Jovian magnetic field.

The most critical phase of mission operations for to study the Galilean satellites of Jupiter occurs at the time of the spacecraft's closest approach to Jupiter ($4\ R_J$). Heavy ions are capable of penetrating the delicate electronics in the spacecraft and causing a stored computer bit to change its value from a "0" to a "1" or vice-versa, a Single Event Upset results (SEU). A single bit flip in one of Galileo's computer memories could trigger a chain reaction of erroneous commands with disastrous results.

Modern microelectronic devices can suffer from single event effects caused by cosmic radiation neutrons in the atmosphere. The phenomenon has been observed both on ground and at aircraft altitudes. The neutron flux at aircraft altitudes (<15 km) is large enough to make the neutron single event effects a problem to aircraft electronics. The most studied device type is static random access memories (SRAM) since those devices have a very high density of transistors, making them sensitive to particle radiation. The cosmic ray neutrons are produced by the charged primary cosmic radiation in the earth's atmosphere. Thereby the atmospheric neutron flux is certainly influenced by solar activity and space weather (see e.g. Dyer, 2001)

Normand (1996) studied the effect of SEU in avionics. Ziegler and Lanford (1979) studied the effect of cosmic rays on computer memories. SEU in implantable cardioverter defibrillators were studied by Bradley and Normand (1998). They found some correlation with the expected geographical variation of the secondary cosmic ray flux.

Prediction of times with increased risk of internal charging on spacecraft are given by Andersson et al. (1999) and Wu et al. (1999).

9.2.6 Solar Activity and Satellite Lifetimes

Satellites in low Earth orbit, with perigee altitudes below 2000 km, are subject to atmospheric drag. This force very slowly circularizes the orbits and the altitude is reduced too. The rate of decay of these orbits becomes extremely rapid at altitudes less than 200 km. As soon as the satellite is down to 180 km it will only have a few hours to live and after several revolutions around the Earth it will re-entry down to Earth. At that phase the temperature is very high and most of the satellite will vaporize. Only large satellites become not fully vaporized and component pieces of them may reach the ground.

The essential parameter for this deceleration is the air density. This varies along the satellite's orbit and is a function of latitude, longitude, time of day, season etc. At a fixed point in space the density can be expressed in terms of the two space environmental parameters:

- 10 cm solar radio flux (F10),
- geomagnetic index A_p.

It is extremely difficult to predict exactly when a satellite will re-enter the atmosphere. The reason for that is that the space environment is not exactly predictable

9.2. SATELLITES

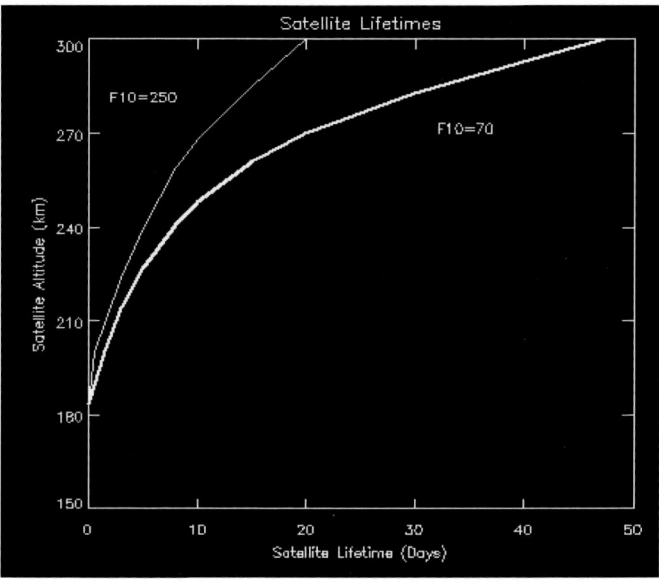

Figure 9.5: Satellite lifetimes

and there are also unresolved variations in atmospheric density. The accuracy of the prediction is in the order of 10 %. That means that one day before re-entry the uncertainty is at least 2 hours. Within that time however, the satellite will have circled the globe and thus it is difficult to predict the location of re-entry with a reasonable warning.

The decay of a satellite's orbit also depends on the cross section of the object itself. In Fig. 9.5 a rough estimate of the lifetime of a satellite with effective mass to cross section ratio 100 kg /m^2 in a circular orbit below 300 km is given for two cases: a) for solar minimum conditions, b) for solar maximum conditions. The geomagnetic field is assumed to be quiet during this period. The lifetime values may be varied for satellites of differing mass to area ratios.

The uncertainty in the predictions is shown by a NORAD prediction in April 1979 for the expected re-entry of the SKYLAB space station between 11 June and 1 July of that year. The actual re-entry occurred on July 11, outside the stated interval, a prediction error from mid-interval of around 15%.

9.2.7 The Atmospheric Model

Let us briefly discuss an atmospheric model that is confined to satellites with orbits totally below 500 km altitude. The reason for that is a simplification: the orbit must then be essentially circular and in place of the orbital radius we can use just the semimajor axis. The atmospheric density ρ is defined by an exponential with variable scale height H. For a fixed exospheric temperature T, H varies with altitude h trough the use of an effective molecular mass m. m includes the actual variation in molecular mass with height and a compensation term for the variation in temperature over the considered range from 180 to 500 km. The variation in density due to the space environment is introduced through T, where $T = T(F10.7\,\text{cm}, A_p)$. Generally, the solar X-ray output incident upon the Earth is absorbed at the base of the thermosphere (120 km) and gives rise to a heating which propagates upward from this level. We use the solar F10.7 cm flux which can vary from 65 to 300 SFU (Solar Flux Units, 1 SFU= 10^{-22} W/m²/Hz) as a proxy for X-rays. The other quantity to take into account is the precipitation of particles- most of them coming from the Sun (CMEs). These are well correlated with the large variations in the geomagnetic field measured at the ground level and quantified by the geomagnetic indices (we use the A_p index here). The A_p is computed every 24 hours and during quiet periods just above zero but may rise up to 400.

Then we can write the following set of equations:

$$T = 900 + 2.5(F10.7 - 70) + 1.5 A_p \quad [\text{Kelvin}] \tag{9.1}$$
$$m = 27 - 0.012(h - 200) \quad 180 < h[\text{km}] < 500 \tag{9.2}$$
$$H = T/m \quad [\text{km}] \tag{9.3}$$
$$\rho = 6 \times 10^{-10} \exp(-(h-175)/H) \quad [\text{kgm}^{-3}] \tag{9.4}$$

The output of this simple model is the density. The intermediate values are only used to derive this density and may not correspond to true atmospheric values at any height within the considered range. The temperature e.g. may be regarded as the mean asymptotic value for the exosphere at large altitudes. The mean molecular weight might be regarded as an integrated mean value from the base of the thermosphere up to the specified height.

The solar 10.7 cm radio flux is used in averaged form (average over the last 90 days). A small correction may be made to weight the current flux more strongly.

Now let us consider the satellite drag. When a spacecraft travels through an atmosphere it experiences a drag force opposite to the direction of its motion. This is given by:

$$D = \frac{1}{2}\rho v^2 A C_d \tag{9.5}$$

D... drag force, ρ atmospheric density, v... speed of the satellite, A... cross sectional area perpendicular to the direction of motion, C_d... drag coefficients. The latter can vary; at altitudes at which satellites orbit $C_d \sim 2$. We introduce the effective cross sectional area $A_e = A C_d$.

For a circular orbit we have the following relation:

$$P^2 G M_e = 4\pi^2 a^3 \tag{9.6}$$

G... gravitational constant, M_e... mass of the Earth. The reduction in the period due to atmospheric drag is given by:

$$\frac{dP}{dt} = -3\pi a \rho \frac{A_e}{m} \tag{9.7}$$

9.2. SATELLITES

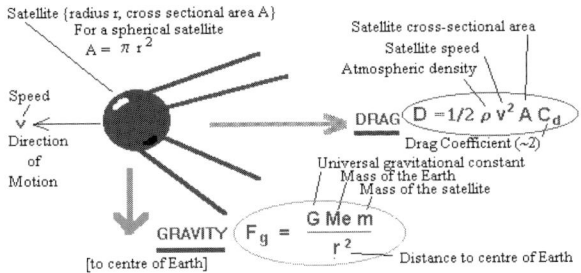

Figure 9.6: Forces acting on a satellite in a low circular orbit.

Re-entry is assumed when the satellite has descended to an altitude of 180 km. The space environmental parameters are given by the solar 10.7 cm radio flux and the geomagnetic activity index. Furthermore one has to provide an estimate for the satellite mass to area ratio. In the absence of any further information this value can be taken as 100 kg /cm^2. This is an average value for many satellites.

If the program underestimates the actual decay of the orbit, you must decrease the mass to area ration, in the case of an overestimation the ratio must be increased.

Also the situation becomes more complicated when considering satellites with very elliptical orbits. Here, a part of the orbit is outside the current atmospheric model. They are also subject to other perturbations (Sun, Moon). If the eccentricity is not too large, one can introduce an effective height in that model:

$$h_{\text{eff}} = q + 900 \times \exp^{0.6} \tag{9.8}$$

q... is the perigee (lowest height) of the orbit and e the eccentricity. For example, the lifetime of a satellite in an elliptical orbit with $e = 0.01$ and $q = 400\,\text{km}$ is the same as the lifetime of a satellite in a circular orbit of height:

$$400 + 900(0.01)^{0.6} \tag{9.9}$$

Since that formula is only a rough approximation it should only be used for orbits with $e < 0.1$. The solar activity should be constant during the orbit decay. Most satellites reaching the end of their lives will have orbits with very low eccentricities (i.e. nearly circular). The reason for this is that atmospheric drag acts to circularize orbits. The apogee height is decreased whilst the perigee height is little affected until the orbit becomes close to circular.

9.2.8 Further Reading

Air Force Geophysics Laboratory, "Handbook of Geophysics and the Space Environment", (1985)

Gatland K, "Space Technology", Salamander Books (London, 1981)

Hargreaves JK, "The Upper Atmosphere and Solar-Terrestrial Relations", Van Nostrand Reinhold (New York, 1979)

King-Hele D, "Observing Earth Satellites", Van-Nostrand Reinhold (New York, 1983)

King-Hele D, "Satellite Orbits in an Atmosphere - Theory and Applications", Blackie, Glasgow (1987)

Tascione T F, "Introduction to the Space Environment", Krieger (Florida, 1994)

Tobiska W K, R D Culp and C A Barth, "Predicted Solar Cycle Twenty-Two 10.7 cm Flux and Satellite Orbit Decay", Journal of the Astronautical Sciences, 1987, pp419-433, vol 35

Reedy (1997) discusses the natural sources of energetic particles in space. The main radiation threats are the galactic cosmic rays (GCRs), solar energetic particles (SEPs) and trapped radiation around planets. Especially outside the Earth's strong magnetosphere, the SEPs are very serious sources of radiation. Over a short period of time (few days) the effects of a huge solar particle event (SPE) ca be greater than any other source of radiation. This causes high doses to humans, microelectronics and solar panels.

An analytical study has been carried out on an impact feature within a solar cell from the Hubble Space telescope Solar array. The feature was investigated optically, and the damage was seen as the result of a partially penetrating impact and therefore some impact particles must have been responsible for that. The residue in the impact was found to contain elements such as Fe, Ti, K, Ca, Si, Mg and Na. The elements Mg, Fe and Ti are usually foreign to a solar cell and this suggests that the impact residue may be of natural or man made origin. Subsequent detailed analysis showed Fe and Mg in concentrations of about 10% and Ti in only limited amounts. That implies that the residue is of natural origin. A more detailed description can be found in Graham et al. (1997)

Chapter 10

The NOAA Space Weather Scales

As we have seen the fast and effective communication of space weather effects to the public is very important. For that reason the US NOAA (National Oceanic and Atmospheric Administration) has introduced the space weather scales. A summary of the different influences triggered by the Sun is shown in Fig. 10.1.

The NOAA space weather scales can be grouped into three different parts:

- Geomagnetic storms
- Solar radiation storms
- Radio blackouts

In the following we will briefly review these scales.

10.1 Geomagnetic Storms

The geomagnetic storms are divided into 5 categories, G1...G5 where the last have the most severe effects.

10.1.1 G1

classified as minor; the influence on power systems is weak, some grid fluctuations can occur. Also the influence on spacecraft is negligible. It seems however that migratory animals are affected even at this low level; the aurora is commonly visible at high latitudes.

As an average about 1700 events per cycle (corresponding to about 900 days per cycle) are to be expected. The K_p value is about 5.

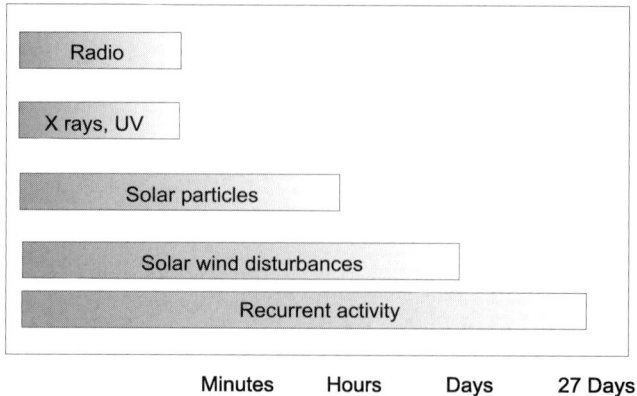

Figure 10.1: After a flare or coronal mass ejection erupts from the Sun's surface, major disturbances arrive with a range of time delays and a storm begins to build in the space surrounding the Earth.

10.1.2 G2

moderate; at this activity some damage may occur in power systems: high-latitude power systems may experience voltage alarms, long-duration storms may cause transformer damage.

Concerning spacecraft operations, corrective actions to orientation may be required by ground control; possible changes in drag affect orbit predictions. This imposes problems to fully automated satellites.

Concerning terrestrial telecommunication the HF radio propagation can fade at higher latitudes, and aurora has been seen as low as New York and Idaho (down to 55 geomagnetic latitude). The K_p value is about 6 and on the average one can expect 600 events per cycle (corresponding to about 360 days per cycle).

10.1.3 G3

strong; on power systems voltage corrections may be required; furthermore false alarms can be triggered on some protection devices.

On satellite components surface charging may occur. Due to the extension of the terrestrial atmosphere during these events drag may increase on low-Earth-orbit satellites, and corrections may be needed for orientation problems.

It is also very important to note that intermittent satellite navigation and low-frequency radio navigation problems may occur, HF radio may be intermittent, and aurora has been seen as low as down to 50 geomagnetic lat. The K_p value is about 7 and on the average one can expect 200 events per cycle (corresponding to 130 days per cycle).

10.1.4 G4

severe; widespread voltage control problems may occur in power systems and some protective systems will mistakenly trip out key assets from the grid.

The problems of surface charging and tracking of satellites increase considerably.

On surface pipelines, induced currents affect preventive measures; the satellite navigation can degrade for hours and the low frequency navigation can be disrupted. Aurora has been seen down to 45^0 geomagnetic latitude. The K_p index is at 8 and on the average one has to count with 100 events per cycle (corresponding to 60 days per cycle).

10.1.5 G5

extreme; widespread voltage control problems and protective system problems can occur; transformers may experience damages and some grid systems may experience complete collapse or blackouts.

The spacecraft operations are affected by extensive surface charging, problems with orientation, uplink/downlink and tracking satellites.

At this activity pipeline currents can reach hundreds of amps, HF radio propagation may be impossible in many areas for one to two days, satellite navigation may be degraded for days, low-frequency radio navigation can be blocked for hours. The aurora has been seen down to 40^0 geomagnetic latitude (Italy, southern Texas). At this level $K_p = 9$ and on the average one has to expect 4 events per cycle (corresponding to about 4 days per cycle).

10.2 Solar Radiation Storms

Again this is a classification from 1 to 5 (S1...S5). This activity can be quantitatively measured by the flux of ions $\geq 10\,\mathrm{MeV}$ as five minute averages in the units $\mathrm{s}^{-1}\mathrm{ster}^{-1}\mathrm{cm}^{-2}$.

10.2.1 S1

minor; there are no effects on biological systems and satellite operations; no danger for astronauts (especially for EVAs (extravehicular activities)). There may be some minor impacts on HF radio in the polar regions. The ion flux is about 10 (see above units). There are about 50 events per solar cycle.

10.2.2 S2

moderate; there are no biological influences; for satellite operations it is important to know that infrequent single-event upsets are possible.

Small effects occur on HF propagation through the polar regions and navigation at polar cap locations is possibly affected.

The ion flux is about 100 and we have about 25 events per cycle.

10.2.3 S3

strong; at this level radiation hazard avoidance is recommended for astronauts on EVA; passengers and crew in commercial jets at high latitudes may receive low-level radiation exposure (equivalent to approximately 1 chest x-ray).

The effects on satellite operations become important: lots of single-event upsets, noise in imaging systems, and slight reduction of efficiency in solar panels are likely.

On Earth, degraded HF radio propagation through the polar regions and navigation position errors are likely. The ion flux is about 10^3 and we have about 10 such events per cycle.

10.2.4 S4

severe; unavoidable radiation hazard to astronauts on EVA thus it is necessary to alarm astronauts; moreover, elevated radiation exposure to passengers and crew in commercial jets at high latitudes (equivalent to approximately 10 chest x-rays) is possible.

Satellites may experience memory device problems and noise on imaging systems; star-tracker problems may cause orientation problems, and solar panel efficiency can be degraded.

On the surface blackout of HF radio communications through the polar regions and increased navigation errors over several days are likely.

The ion flux is about 10^4. There are about 3 such events per cycle.

10.2.5 S5

extreme; unavoidable high radiation hazard to astronauts on EVA; high radiation exposure to passengers and crew in commercial jets at high latitudes (equivalent to approximately 100 chest x-rays) is possible.

Satellites may be put out of operation, memory impacts can cause loss of control, may cause serious noise in image data, star-trackers may be unable to locate sources; permanent damage to solar panels is possible.

At the surface complete blackout of HF communications is possible through the polar regions, and position errors make navigation operations extremely difficult.

The ion flux is at 10^5; fortunately, these events occur on a rate fewer than 1 per cycle.

10.3 Scale for Radio Blackouts

Classified as R1...R5; measured as GOES X-ray peak brightness by class, measured in the 0.1-0.8 nm range, in Wm^{-2}.

10.3. SCALE FOR RADIO BLACKOUTS

10.3.1 R1

minor; we have to take into account a weak or minor degradation of HF radio communication on the sunlit side, as well as occasional loss of radio contact.

Concerning navigation we have to consider that low-frequency navigation signals may be degraded for brief intervals. The physical measurement is M1 and (10^{-5}). On the average 2000 such perturbances per cycle occur (on 950 days per cycle).

10.3.2 R2

moderate; limited blackout of HF radio communication on sunlit side occur, loss of radio contact for tens of minutes.

Navigation: a degradation of low-frequency navigation signals for tens of minutes is likely. The physical classification of the relevant solar event goes M5 and the flux to 5×10^{-5}.

On the average one has 350 events per cycle (300 days per cycle).

10.3.3 R3

strong; a wide area blackout of HF radio communication, as well as a loss of radio contact for about an hour on the sunlit side of Earth is likely.

Since low-frequency navigation signals are being degraded for about an hour this also has serious consequences for navigation.

The physical classification is X1, the flux 10^{-4} and one has 175 events per cycle (140 days per cycle).

10.3.4 R4

severe; HF radio communication blackout occurs mostly on the sunlit side of Earth for one to two hours and a HF radio contact loss during this time has to be expected.

Outages of low-frequency navigation signals cause increased error in positioning of navigational systems for one to two hours. Minor disruptions of satellite navigation are likely on the sunlit side of Earth.

The physical classification is X10, the flux 10^{-3} and one has 8 events per cycle (8 days per cycle).

10.3.5 R5

extreme; a Complete HF (high frequency) radio blackout on the entire sunlit side of the Earth lasting for a number of hours may occur . This results in no HF radio contact with mariners and en route aviators in this sector.

Navigation: Low-frequency navigation signals used by maritime and general aviation systems experience outages on the sunlit side of the Earth for many hours, causing loss in positioning. Satellite navigation errors in positioning increase for several hours on the sunlit side of Earth, which may spread into the night side.

The physical classification is X20, the flux 2×10^{-3} and one has less than 1 events per cycle.

10.4 Summary

The classification scheme given above enables very easily to estimate the effect of geomagnetic storms and solar radiation storms on satellites and telecommunication systems. This is also extremely important for manned space mission (ISS, international space station). On the other hand, the solar activity is declining again after heaving reached its maximum in 2000. One can estimate that there will be about 25 EVA/year necessary for the construction of the space station. For that reason, it is extremely important to alert astronauts for S4 and S5 storms. The predicted sales figures for GPS systems rise from 5000 Million USD for 1998 and more than 9000 Million USD for 2000. This means that more and more systems are equipped with these navigation systems but on the other hand we must take into account that small degradations may even occur at R1 levels. The frequency of such events is however more than 2000 per cycle.

Also the number of satellites will increase.

Skylab is an example of a spacecraft re-entering Earth's atmosphere prematurely as a result of higher-than-expected solar activity because of the increased drag.

Systems such as LORAN and OMEGA are adversely affected when solar activity disrupts their radio wavelengths. The OMEGA system consists of eight transmitters located through out the world. Airplanes and ships use the very low frequency signals from these transmitters to determine their positions. During solar events and geomagnetic storms, the system can give navigators information that is inaccurate by as much as several miles. If navigators are alerted that a proton event or geomagnetic storm is in progress, they can switch to a backup system. GPS signals are affected when solar activity causes sudden variations in the density of the ionosphere.

We have seen before that some military detection or early-warning systems are also affected by solar activity. The Over-the-Horizon Radar bounces signals off the ionosphere in order to monitor the launch of aircraft and missiles from long distances. During geomagnetic storms, this system can be severely hampered by radio clutter. That can occur at even low activity (R1 perturbances).

10.5 Space Weather on Mars

Because Mars will be a target of future manned space missions we briefly discuss space weather influences there.

Mars is a completely different world. It is dry like a desert, cold as the Earth's Antarctic and possibly lifeless. Future human colonists will have a different set of weather conditions. The Earth is protected by the magnetosphere. Mars does not possess a global magnetic field to shield it from solar flares and cosmic rays. Scientists aren't sure why, but Mars' internal magnetic dynamo turned off about

10.5. SPACE WEATHER ON MARS

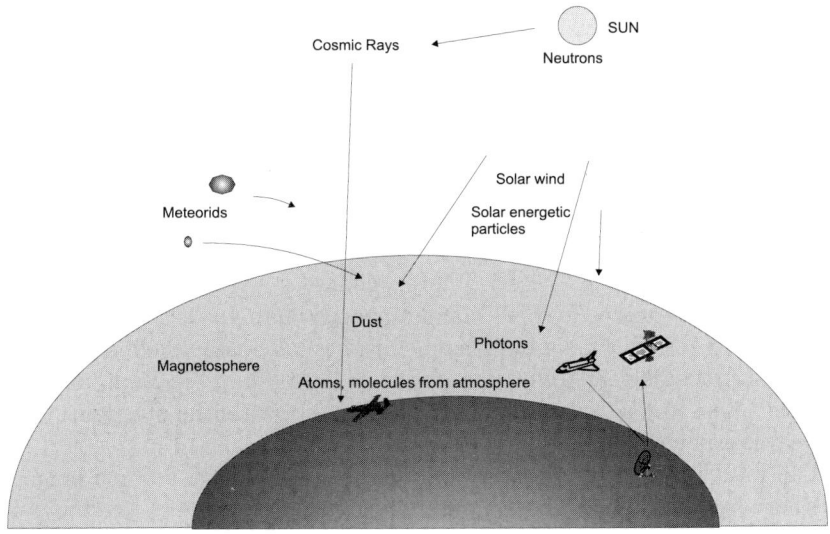

Figure 10.2: Summary of space weather effect in the Earth's environment

4 billion years ago. After that, the solar wind gradually eroded the martian atmosphere until, today, it has less than 1% of the thickness of the Earth's.

No global magnetic field and a very thin atmosphere – those are the two factors that render Mars vulnerable to space radiation. Does that mean that because of this unprotected exposure Mars is lifeless? It is assumed that certain life forms could be radiation resistant like the terrestrial microbe Deinococcus radiodurans. D. radiodurans has a feature that is considered all-important in aerospace: redundancy. Its genetic code repeats itself many times so that damage in one area can be recognized and repaired quickly. It withstands attacks from acid baths, high and low temperatures, and even radiation doses, e.g. the microbe can withstand without loss of viability a dosage that is 3 000 times greater than what would kill a human.

Tiny Martians might also live in rocks or soil, substances that provide natural protection against radiation. In addition to that, magnetic fields provide some protection, ancient global, regional and sub surface fields.

NASA sent a radiation monitor, MARIE (Mars Radiation Environment Experiment) to the Red Planet on April 7th with the 2001 Mars Odyssey spacecraft. MARIE is one of three scientific instruments on board – the other two will search for signs of water and interesting minerals on Mars. MARIE can detect charged particles (electrons, protons, ions) with energies between 15 MeV and 500 MeV.

As we have seen in the previous chapter, particulate radiation poses the greater threat to humans. Particles from solar flares (protons) are of greater concern - here particles with relatively low energies (around 70 MeV) are produced. Such protons lose energy in tissue. Cosmic ray nuclei have energies between 300 and 500 MeV and they penetrate the human body in a short time so that there is not enough time to transfer their energy to the surrounding tissue. Solar protons when passing through humans ionize molecules along their track- the ionization creates free radicals causing modification or even break of the DNA strands and if the cell survives it can become cancerous.

Therefore, human settlers on Mars must be protected from these energetic protons. The air density at the surface of Mars is equivalent to that of the Earth's atmosphere at 20 km altitude. Astronauts must be protected by shelter walls.

Since Mars has no global magnetic field the surface is eroded by the solar wind as well as by the planet's atmosphere. Currently Mars looses approximately 2 kg/s of its atmosphere. In 1998 magnetometers discovered a network of magnetic loops arrayed across Mars's southern hemisphere. Locally, the magnetic fields arch over the surface like umbrellas, hundreds of km high. In such an area you would measure a field about as strong as the Earth's (a few tenths of a gauss). Elsewhere the field is extremely weak.

The martian ionosphere traces the distribution of the surface magnetic field, and there seems to be a 1-to-1 correspondence: places where magnetic umbrellas deflect the solar wind are also spots where the ionosphere is retained high above the surface of the planet.

The Earth's global magnetic field is caused and maintained from an active dynamo – that is, circulating currents at the planet's liquid metallic core. A similar dynamo once churned inside Mars, but for reasons unknown it stopped working four billion years ago. The patchwork fields mentioned above, we see now are remnants of that original magnetic field. Hellas and Argyre, two large impact basins on Mars are about four billion years old and are demagnetized. If the dynamo was still operating when those impact features formed, the crust would have re-magnetized as they cooled. Hence, the dynamo must have stopped before then.

Chapter 11

Asteroids, Comets, Meteroites

11.1 Asteroids

11.1.1 General Properties

On the first day of January 1801, Giuseppe Piazzi discovered an object which he first thought was a new comet. But after its orbit was better determined it was clear that it was not a comet but more like a small planet and it was therefore named asteroid. The proper name of the first asteroid detected is Ceres. Three other small bodies were discovered in the next few years (Pallas, Vesta, and Juno). By the end of the 19th century several hundred asteroids were known.

Several thousand asteroids have been discovered and given provisional designations so far. Thousands more are discovered each year. There are undoubtedly hundreds of thousands more that are too small to be seen from the Earth. There are 26 known asteroids larger than 200 km in diameter. About 99 % of all objects > 100 km are known however of the total number of asteroids with diameters between 10 and 100 km we know only 50%. It is difficult to estimate the total number of asteroids, perhaps as many as a million 1 km sized asteroids may exist.

Since most of the asteroids have orbits between Jupiter and Mars, it was first assumed that they are remnants of a larger planet that broke up. However, the total mass of all the asteroids is less than that of the Moon.

Ceres has a diameter of 933 km, the next largest are Pallas, Vesta and Hygiea which are between 400 and 525 km in diameter. All other known asteroids are less than 340 km.

Asteroids are classified into:

- C-type: extremely dark (albedo 0.03), similar to carbonaceous chondrite meteorites; 75% of known asteroids.

- S-type: 17% of asteroids; bright (albedo 0.1-0.2); metallic Ni, Fe and Mg silicates.

- M-type: bright (albedo 0.1-0.2), pure NiFe.

- rare types

One should however take into account biases in the observations- e.g. dark C-types are more difficult to detect. According to their position in the solar system, asteroids can also be categorized into:

- Main belt: located between Mars and Jupiter, 2-4 AU from the Sun.

- Near Earth Asteroids (NEAs): they closely approach the Earth and will be treated separately.

- Trojans: located near Jupiter's Lagrange points (60 degrees ahead and behind Jupiter in its orbit); several 100 are known.

- Between the main concentration in the Main Belt are relatively empty regions known as Kirkwood gaps. These are regions were an object's orbital period would be a simple fraction of that of Jupiter (resonance).

- Centaurs: asteroids in the outer solar system; e.g. Chiron (his orbit lies between Saturn and Uranus).

Oort proposed in 1950, that comets reside in a vast cloud at the outer reaches of the solar system. This has come to be known as the Oort Cloud. This hypothesis is based on several observational facts: a) no comet has been observed with an hyperbolic orbit (which would indicate interstellar origin), b) aphelia of long period comets lie at a distance of about 50 000 AU, c) there is no preferential direction from which comets come. The Oort cloud may contain up to 10^{12} comets (in total about the mass of Jupiter). The Kuiper Belt is a disk-shaped region past the orbit of Neptune roughly 30 to 100 AU from the Sun containing many small icy bodies. It is now considered to be the source of the short-period comets.

11.2 Potentially Hazardous Asteroids

Potentially Hazardous Asteroids (PHAs) are currently defined based on parameters that measure the asteroid's potential to make threatening close approaches to the Earth.

To be classified as PHA, the following parameters must be fulfilled:

- an Earth Minimum Orbit Intersection Distance (MOID) of 0.05 AU or less,

- absolute magnitudes (H) of 22.0 or less are considered.

An asteroid's absolute magnitude H is the visual magnitude an observer would record if the asteroid were placed 1 Astronomical Unit (AU) away, and 1 AU from the Sun and at a zero phase angle. The diameter of an asteroid can estimated from its absolute magnitude (H). The lower the H value, the larger the size of the object. However, this also requires that the asteroid's albedo be known as well.

Since the albedo for most asteroids is not known, an albedo range between 0.25 to 0.05 is usually assumed. This results in a range for the diameter of the asteroid. The table 11.4 shows the diameter ranges for an asteroid based on its absolute magnitude, assuming an albedo ranging from 0.25 to 0.05.

In other words, asteroids that can't get any closer to the Earth (i.e. MOID) than 0.05 AU (roughly 7,480,000 km) or are smaller than about 150 m in diameter (i.e. H = 22.0 with assumed albedo of 13%) are not considered PHAs. The current list of PHAs is obtained from the Minor Planet Center on a daily basis. Asteroids with a small MOID to Earth should be carefully followed because they can become Earth colliders.

Because of long-range planetary gravitational perturbations and, particularly, close planetary approaches, asteroid orbits change with time. Consequently, MOID also changes. As a rule of thumb, MOID can change by up to 0.02 AU per century, except for approaches within 1 AU of massive Jupiter, where the change can be larger. Thus, an asteroid that has a small MOID with any planet should be monitored. Currently there are about 350 known PHA's.

11.3 Torino Impact Scale

This was established (analogous to the space weather scale) to characterize different objects.

Events Having No Likely Consequences (White Zone)

0 The likelihood of a collision is zero, or well below the chance that a random object of the same size will strike the Earth within the next few decades. This designation also applies to any small object that, in the event of a collision, is unlikely to reach the Earth's surface intact.

Events Meriting Careful Monitoring (Green Zone)

1 The chance of collision is extremely unlikely, about the same as a random object of the same size striking the Earth within the next few decades.

Events Meriting Concern (Yellow Zone)

2 A somewhat close, but not unusual encounter. Collision is very unlikely. **3** A close encounter, with 1% or greater chance of a collision capable of causing localized destruction. **4** A close encounter, with 1% or greater chance of a collision capable of causing regional devastation.

Threatening Events (Orange Zone)

5 A close encounter, with a significant threat of a collision capable of causing regional devastation. **6** A close encounter, with a significant threat of a collision capable of causing a global catastrophe. **7** A close encounter, with an extremely significant threat of a collision capable of causing a global catastrophe.

Group	Description	Definition
NECs	Near-Earth Comets	q<1.3 AU, P<200 years
NEAs	Near-Earth Asteroids	q< 1.3 AU
Atens	Earth-crossing NEAs	a<1.0 AU, Q>0.983 AU
Apollos	Earth-crossing NEAs	a>1.0 AU, q<1.017 AU
Amors	Earth-appr. NEAs with orbits betw. Earth and Mars	a>1.0 AU, 1.017<q<1.3 AU
PHAs	NEAs	MOID≤0.05 AU, H≤22.0

Certain Collisions (Red Zone)

8 A collision capable of causing localized destruction. Such events occur somewhere on Earth between once per 50 years and once per 1 000 years. **9** A collision capable of causing regional devastation. Such events occur between once per 1 000 years and once per 100 000 years. **10** A collision capable of causing a global climatic catastrophe. Such events occur once per 100 000 years, or less often.

11.4 NEOs

Near-Earth Objects (NEOs) are comets and asteroids that have been nudged by the gravitational attraction of nearby planets into orbits that allow them to enter the Earth's neighborhood. Composed mostly of water ice with embedded dust particles, comets originally formed in the cold outer planetary system while most of the rocky asteroids formed in the warmer inner solar system between the orbits of Mars and Jupiter. The scientific interest in comets and asteroids is due largely to their status as the relatively unchanged remnant debris from the solar system formation process some 4.6 billion years ago. The giant outer planets (Jupiter, Saturn, Uranus, and Neptune) formed from an agglomeration of billions of comets and the left over bits and pieces from this formation process are the comets we see today. Likewise, today's asteroids are the bits and pieces left over from the initial agglomeration of the inner planets that include Mercury, Venus, Earth, and Mars.

As the primitive, leftover building blocks of the solar system formation process, comets and asteroids offer clues to the chemical mixture from which the planets formed some 4.6 billion years ago. If we wish to know the composition of the primordial mixture from which the planets formed, then we must determine the chemical constituents of the leftover debris from this formation process - the comets and asteroids.

In terms of orbital elements, NEOs are asteroids and comets with perihelion distance q less than 1.3 AU. Near-Earth Comets (NECs) are further restricted to include only short-period comets (i.e orbital period P less than 200 years). The vast majority of NEOs are asteroids, referred to as Near-Earth Asteroids (NEAs). NEAs are divided into groups (Aten, Apollo, Amor) according to their perihelion distance (q), aphelion distance (Q) and their semi-major axes (a).

Possible NEO missions that require spacecraft with the capability to ren-

absolute Magn.	Diameter	absolute Magn.	Diameter
3.0	670 km - 1490 km	3.5	530 km - 1190 km
4.0	420 km - 940 km	4.5	330 km - 750 km
5.0	270 km - 590 km	5.5	210 km - 470 km
6.0	170 km - 380 km	6.5	130 km - 300 km
7.0	110 km - 240 km	7.5	85 km - 190 km
8.0	65 km - 150 km	8.5	50 km - 120 km
9.0	40 km - 90 km	9.5	35 km - 75 km
10.0	25 km - 60 km	10.5	20 km - 50 km
11.0	15 km - 40 km	11.5	13 km - 30 km
12.0	11 km - 24 km	12.5	8 km - 19 km
13.0	7 km - 15 km	13.5	5 km - 12 km
14.0	4 km - 9 km	14.5	3 km - 7 km
15.0	3 km - 6 km	15.5	2 km - 5 km
16.0	2 km - 4 km	16.5	1 km - 3 km
17.0	1 km - 2 km	17.5	1 km - 2 km
18.0	670 m - 1500 m	18.5	530 m - 1200 m
19.0	420 m - 940 m	19.5	n330 m - 750 m
20.0	270 m - 590 m	20.5	210 m - 470 m
21.0	170 m - 380 m	21.5	130 m - 300 m
22.0	110 m - 240 m	22.5	85 m - 190 m
23.0	65 m - 150 m	23.5	50 m - 120 m
24.0	40 m - 95 m	24.5	35 m - 75 m
25.0	25 m - 60 m	25.5	20 m - 50 m
26.0	17 m - 37 m	26.5	13 m - 30 m
27.0	11 m - 24 m	27.5	8 m - 19 m
28.0	7 m - 15 m	28.5	5 m - 12 m

dezvous at great distances (1 AU) from the Earth within a releatively short amount of time (on the order of a year) are discussed by Sforza and Remo (1997) and Powell et al. (1997). NEOs as near Earth resources for mining are discussed by Gertsch et al. (1997).

11.5 The Cretaceous-Tertiary Impact

65 million years ago at the Cretaceous-Tertiary boundary (K/T) an impact occurred. Mass extinctions of a broad spectrum of lifeforms (Raup, 1989), a worldwide clay layer containing geochemical (Alvarez et al., 1980), mineralogical (Bohor , 1990) and isotopic anomalies (McDougall, 1988) and tsunami deposits (Bourgeois et al., 1988) point to a major event at that time.

The buried Chicxulub basin is the source crater about 300 km in diameter. It is believed that the Chicxulub crater would most likely be formed by a long-period comet composed primarily of nonsilicate materials (ice, hydrocarbons etc.) and subordinate amounts primitive chondritic material. The collision would have raised the energy equivalent to between 4×10^8 and 4×10^9 megatons of TNT. Studies of terrestrial impact rates suggest that such an event would have a mean production rate of $\sim 1.25 \times 10^{-9}\,\text{yr}^{-1}$. This rate is considerably lower that that of the major mass extinctions over the last 250 million years ($\sim 5 \times 10^{-7}\,\text{yr}^{-1}$). However, there is substantial evidence establishing the cause-link between the Chicxulub basin forming event and the K/T biological extinctions. The crater showed several rings (similar to the rings of the Mare Orientale on the Moon).

Let us consider the impact of a small asteroid. It fragments in the atmosphere thus the cross section for aerodynamic braking is greatly enhanced. Ground impact damage such as

- craters,
- earthquakes,
- tsunami

from a stone asteroid is negligible if it is less than 200 m in diameter. Small and relatively frequent impactors such as Tunguska produce only air blast damage and leave no long term scars. Objects 2.5 times larger which hit every few thousand years cause coherent destruction over many thousand km of coast. Let us assume an asteroid > 200 m hits an ocean. A water wave generated by such an impactor has a long range because it is two-dimensional; its height falls off inversely with distance from the impact. When the wave strikes a continental shelf, its speed decreases and its height increases to produce tsunamis. Tsunamis produce most of the damage from asteroids between 200 m and 1 km. An impact anywhere in the Atlantic by an asteroid 400 m in diameter would devastate the coasts on both sides of the ocean by tsunamis over 100 m high. An asteroid 5 km in diameter hitting the mid Atlantic would produce tsunami that would inundate the entire East coast of the US to the Appalachian mountains (see the paper of Hill and Mader, 1997).

In Fig.11.1 the estimated frequency of impacts as a function of asteroid diameter is shown.

An asteroid or comet $\geq 10\,\text{km}$ in diameter (which releases $\geq 10^{24}\,\text{J}$ or 10^8 Mt TNT) would cause a global catastrophe:

- production of dense clouds of ejecta,
- smoke clouds,
- large amounts of nitric oxides are created $\rightarrow$ acid rain;
- the NO_x in the stratosphere destroys the ozone layer.
- in an ocean impact: enhanced greenhouse effect from water vapor injected into the atmosphere,
- CO_2 released by impact into carbonate rocks...

11.5. THE CRETACEOUS-TERTIARY IMPACT

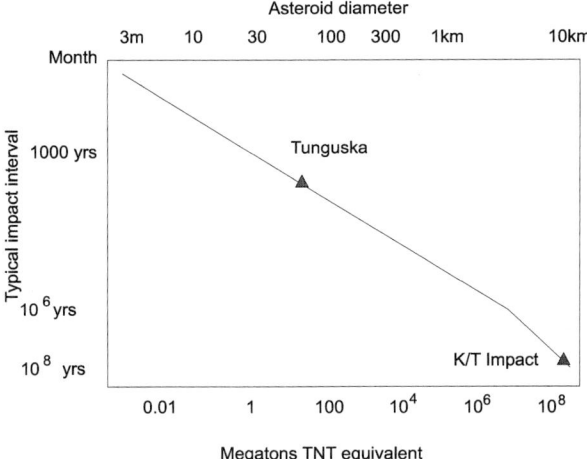

Figure 11.1: Estimated frequency of impacts on the Earth from the present population of comets and asteroids and impact craters.

Some more details about such scenarios can be found e.g. in Chapman and Morrison (1994) or Melosh et al. (1990).

During the last 500 Million years there occurred several extinctions of marine species: in Table 11.1 we give the formal end of stage in Myr.

From the data we can deduce that the Earth should be hit by several asteroids and comets larger than a few km ($\sim 10^{23}$ J energy release) and perhaps one $\leq$10 km in a period of $\sim$ 100 Myr. Thus for the last 500 Myr years 5 events of extinctions are to be expected and 20 minor events which is in agreement with astronomical predictions.

From normal meteoroid ablation an iridium anomaly is observed and is one of the most significant signatures for impact. The search for iridium has resulted in reports of elevated iridium levels ($\leq$ 10 times background values) at or near a number of extinction boundaries. The Ir levels are generally significantly weaker than the K/T anomaly.

The conversion from projectile mass to crater size was given by Shoemaker et al. (1990):

$$D = 0.074 C_f (g_e/g)^{1/6} (W \rho_a/\rho_t)^{1/3.4} \quad (11.1)$$

D...crater diameter, c_f...crater collapse factor (1.3 for craters larger than 4 km on Earth), g_e...gravitational acceleration at the surface of the Earth, g... acceleration at the surface of the body on which the crater is formed (in this case Earth), W... the kinetic energy of the impacting body in kilotons TNT, ρ_a...density of the impactor (1.8 g/cm^3 for a comet...7.3 g/cm^3 for an iron body) and ρ_t...density of the target rock (e.g. 2.7 g/cm^3). From this equation we see, that a carbonaceous chondrite would produce a crater $\sim$ 94 km in diameter whereas an impactor of cometary composition $\sim$ 150 km.

Table 11.1: Extinction of marine species. The end of the stage is given in Myr.

Name	End	Name	End
Pliocene	2.3	Mid Miocene	11.2
Upper Eocene	34	Maastrichtian	65
Cenomanian	93.5	Aptian	112
Tithonian	144	Callovian	159
Pliensbachian	190	Norian	206
Carnian	221	Tatarian	248
Guadelupian	250	Stephanian	290
Serpukhovian	322	Famennian	363
Frasnian	367	Eifelian	380
Ludlovian	411	Ashgillian	439
Llanvirnian	469	Tremadocian	493
Trempeleauean	505	Franconian	508
Botomian	520	Proterozoic/Cambrian	540

11.6 Meteorites

The term meteor comes from the Greek meteoron, meaning phenomenon in the sky. Meteors are small solid particles that enter the Earth's atmosphere from interplanetary space. They move at high speeds and the friction they encounter in the air vaporizes them (typically at heights between 80 and 110 km above the surface). The light caused by the luminous vapors formed in such an encounter appears like a star moving rapidly across the sky, fading within a few seconds. To be visible, a meteor must be within 200 km of the observer. The total number of meteors bright enough to be visible is estimated to be about 25 million per day.

A meteoroid is matter revolving around the sun or any object in interplanetary space that is too small to be called an asteroid or a comet. Even smaller particles are called micrometeoroids or cosmic dust grains, which includes any interstellar material that should happen to enter our solar system. A meteorite is a meteoroid that reaches the surface of the Earth without being completely vaporized.

One of the primary goals of studying meteorites is to determine the history and origin of their parent bodies. Several achondrites sampled from Antarctica since 1981 have conclusively been shown to have originated from the moon based on compositional matches of lunar rocks obtained by the Apollo missions of 1969-1972. Sources of other specific metorites remain unproven, although another set of eight achondrites are suspected to have come from Mars. These meteorites contain atmospheric gases trapped in shock melted minerals which match the composition of the Martian atmosphere as measured by the Viking landers in 1976. All other groups are presumed to have originated on asteroids or comets; the majority of meteorites are believed to be fragments of asteroids.

A typical bright meteor is produced by a particle with a mass less than 1 g. A particle the size of a golf ball produces a bright fireball. The total mass of meteoritic material entering the Earth's atmosphere is estimated to be about 100

11.6. METEORITES

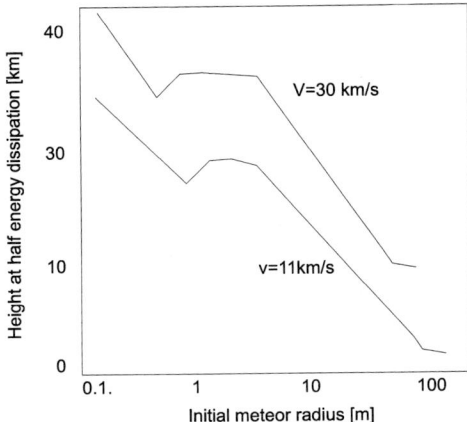

Figure 11.2: Height in the atmosphere at which half the kinetic energy of a stony meteoroid is dissipated. Note that asteroids with > 100 m hit the ground with most of their original kinetic energy.

tons per day.

Meteorites can be classified into stony, stony iron and iron. The most common meteorites are chondrites which are stony. Radiometric dating indicate an age of about 4.5×10^9 years. Achondrites are also stony but they are considered differentiated or reprocessed matter. They are formed by melting and recrystallization on or within meteorite parent bodies. Pallasites are stony iron meteorites composed of olivine enclosed in metal.

The motion of meteoroids can be severely perturbed by the gravitational fields of major planets. Jupiter's gravitational influence is capable of reshaping an asteroid's orbit from the main belt so that it dives into the inner solar system and crosses the orbit of Earth. This is apparently the case of the Apollo and Vesta asteroid fragments.

Particles found in highly correlated orbits are called a stream components and those found in random orbits are called sporadic components. It is thought that most meteor streams are formed by the decay of a comet nucleus and consequently are spread around the original orbit of the comet. When Earth's orbit intersects a meteor stream, the meteor rate is increased and a meteor shower results. Typically, a meteor shower will be active for several days. A particularly intense meteor shower is called a meteor storm. Sporadic meteors are believed to have had a gradual loss of orbital coherence with a meteor shower due to collisions and radiative effects, further enhanced by gravitational influences. There is still some debate concerning sporadic meteors and their relationship with showers. A well known meteor shower are the Perseids (named after the radiant that is the name of the constellation where the meteorites are coming from) which has its maximum on August 11.

11.6.1 The Leonid Threat

On Nov. 18th the Earth glides into a dust cloud shed by the comet Tempel-Tuttle in 1766. Around this date lots of meteors can be seen. The Leonid meteor shower is seen to emanate from a point in the western part of Leo which is called the radiant. When Leonids rain down on the airless moon, they will not cause the shooting stars because there is no atmosphere; they just hit the ground. In 1999 while the Moon passed through the Leonid debris, impact flashes were recorded. When such a meteoroid hits the Moon it vaporizes some dust and rock. Some of those vapors contain sodium (constituent of lunar rocks) which scatters sunlight. The Moon is also surrounded by a gaseous halo called the lunar exosphere ($\sim$ 100 atoms per cm^3). The solar wind blows it into a long tail (much like a comet tail). This tail points away from the Sun and extends several 10^5 km and the Earth passes through it once a month around the time of New Moon. Using extraordinary sensitive cameras, sunlight scattered can be detected. After the Leonid fireball shower of 1998 the density of the Moon's sodium trail tripled.

There is also a risk to satellites. The damage can be:

- mechanical: direct impact through spalling or chipping as larger particles hit the spacecraft and break up;

- electrical: electrostatic discharges (ESDs), electromagnetic pulse (EMP).

Mechanical damage consists predominantly of sandblasting which all spacecraft experience during the Leonids. This causes surface degradations. This is in general not a serious problem. Impacts and spall result from larger particles hitting a satellite. These can punch holes in a solar panel or wall. Spall produced by secondary particles can affect the internal mechanisms of the spacecraft more seriously than the original impact.

EMPs are created from the direct vaporization of impacting particles into plasma and ESDs by a buildup of charging over the satellite surface. Both can cause electrical and communications problems, erroneous signals in telemetry and short circuits.

How can such problems be minimized? One simple manoeuvre is to minimize the cross-sectional area of the satellite that is exposed to the meteor shower; e.g. the solar panels point edge on into the meteor stream by reorienting the spacecraft. Another technique is to turn off equipment that is particularly sensitive to ESDs.

USAF perspectives on Leonid threat are discussed by Treu et al. (1998). Meteoroid impacts on spacecrafts and penetration damage are studied by McBride and McDonnell (1999). Meteoroid morphology and density (e.g. using NASA's LDEF satellite results) were investigated by McDonnel and Gardner (1998).

Chapter 12

Space Debris

Orbital debris is defined as any man-made object in orbit around the Earth which no longer serves as a useful purpose.

In 1957 Sputnik 1 was launched as the first man made spacecraft. In the years of space activities some 3 750 launches led to more than 23 000 observable space objects (larger than 10 cm) of which currently 7 500 are still in orbit. Only 6% of the catalogued orbit population comprise operational spacecraft, while 50% can be attributed to decommissioned satellites, spent upper stages, and mission related objects (launch adapters, lens covers, etc.). The remainder of 44% is originating from 129 on-orbit fragmentations which have been recorded since 1961. These events, all but 1 or 2 of them explosions of spacecraft and upper stages, are assumed to have generated a population of objects larger than 1 cm on the order of 70 000 to 120 000. Only at sizes of in the range of 0.1 mm the sporadic flux from meteoroids prevails over man-made debris. From a statistical point of view we have to note that most orbital debris reside within 2 000 km of the Earth's surface. Within this volume, the amount of debris varies significantly with altitude and regions of debris concentration are found near 800 km, 1 000 km and 1 500 km.

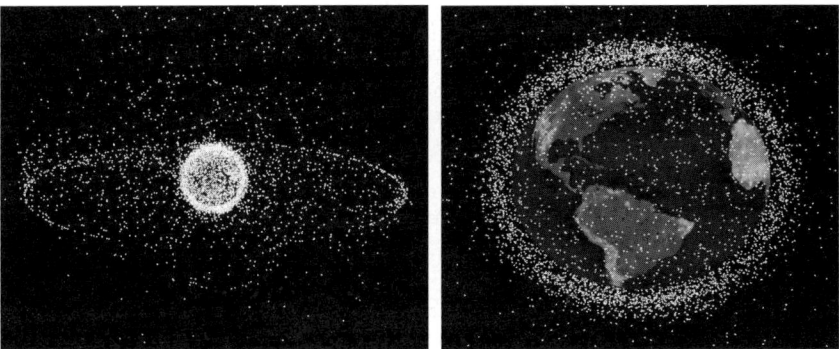

Figure 12.1: GEO and LEO objects as a source of space debris

Space debris is an inherently international problem and its solution requires international co-operation. The Inter-Agency Space Debris Coordination Committee (IADC) whose members are ESA, NASDA (Japan), NASA, and the Russian Space Agency RKA and the Canadian Space Agency (CSA) provides a forum for discussion and coordination of technical space debris issues.

To assess the risk potential of collisions of man-made or natural particulates with operational spacecraft, one must refer to statistical models of the particle population for all size regimes except for man-made debris above 10 cm. In the latter case, collision events or near-miss events can be predicted on the basis of orbital data from operational surveillance networks of the USA or of the CIS. In the former case, collision fluxes can only be estimated statistically. Currently, space debris between 1 cm and 10 cm are neither observable, nor are they shieldable with available on-orbit technology. Hyper-Velocity Impact (HVI) tests are used to experimentally verify and improve shields for on-orbit use (e.g. for Space Station Alpha), with the aim to increase the shieldable impactor size beyond 1 cm. The Mission Analysis Section of ESOC is coordinating all Space Debris Research Activities within ESA. SOC's Meteoroid and Space Debris Terrestrial Environment Reference (MASTER) model can be used to assess the debris or meteoroid flux imparted on a spacecraft on an arbitrary earth orbit.

On Dec 3, 2001 BBC reports, that space debris lit up the sky. The spectacular nighttime light show was seen over parts of southern England is now believed to have been caused by burning Russian space debris. Observers said the fragments, which could be seen over parts of Essex and Sussex, were very bright and traced across the sky for up to four minutes.

At NASA, a new modelling technique called Smooth Particle Hydrodynamics (SPH) is under development. Their approach models the distribution of debris fragments from a collision without using the normal computational mesh that is often subject to tangling. SPH eliminates many difficulties of previous calculation techniques.

From the above considerations it is clear that spacecrafts have to be protected from collisions with space debris. Let us mention two examples: the US space command examines the trajectories of the Space Shuttle in order to identify possible close encounters with space debris. If a dangerous object is believed to approach a few tens of kilometers to the Space Shuttle, it will be maneuvered away from the object (although in such a case the chances of a collision are only approximately 1:100 000). Such an operation is necessary about once every year or two (at present).

Of course the International Space Station (ISS) will be the most heavily shielded spacecraft ever flown. Critical components (e.g., habitable compartments and high press tanks) will normally be able to withstand the impact of debris as large as 1 cm in diameter. ISS will also have manoeuvering capability to avoid hazardous objects.

12.1 Reentry of Orbital Debris

How long will orbital debris remain in Earth orbit? As a rule of thumb one can say that the higher the altitude, the longer the orbital debris will typically remain in Earth orbit.

- Debris left in orbits below 600 km: normally falls back to Earth within a few years.

- Debris left in orbits at altitudes of 800 km: the time for orbital decay is several decades.

- Debris left in orbits at altitudes above 1 000 km: will normally continue circling the Earth for a century or more.

Up to now no serious injury or property damage has been confirmed caused by reentering debris. Most of the space debris does not survive the severe heating which occurs during reentry. Components fall most likely into the oceans or onto sparsely populated regions like the Canadian Tundra (in that case a contamination with Plutonium occurred) or Siberia. During the past 40 years, on the average one cataloged piece of debris fell back to earth each day.

On 12 June 1979 Skylab came crashing to Earth, scattering chunks of metal over the West Australian desert. US officials were unable to control it's final descent. Pieces of the Russian space station Mir race across the sky above Fiji as it makes its descent into the earth's atmosphere on March 23, 2001. Mir plunged to earth after Russian Mission Control fired engines to nudge it out of the orbit it has kept for 15 years. The entrance velocity was 6 400 km/h and the final burst of rockets was made at a height of only 170 km over Africa. The weight of the space station was about 135 tons.

12.2 Orbital Debris Protection

Many efforts are made to develop protection:

- hypervelocity impact measurements: in such experiments projectiles are produced at speeds more than seven times faster than the fastest bullet; this is done with so called two stage light gas guns. The impact event last only a few microseconds. The velocity of the bullet is measured by using two laser curtains positioned a short distance uprange of the target. The distance between the curtains is known and the time elapsed between the two disruptions is measured, thus the projectile velocity can be measured.

- Shield development.

- Simulations: sophisticated computer programs simulating hypervelocity events are run on supercomputers. This approach to developing spacecraft shield solutions is becoming more and more prevalent.

- Developing new materials

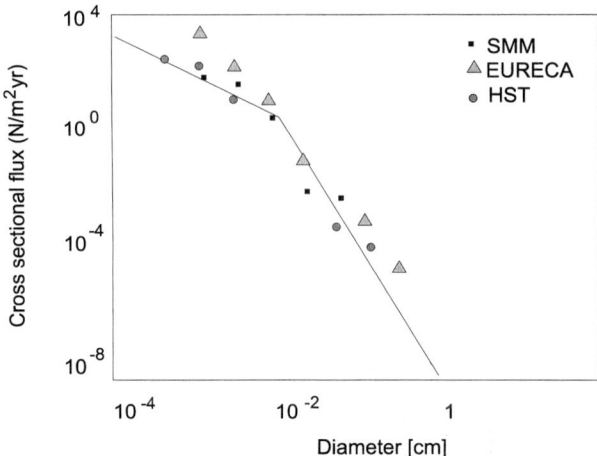

Figure 12.2: Approximate measured debris flux in low Earth orbit by object size (sketch)

- Impacts on spacecraft: all spacecraft collide with very small orbital debris particles and meteoroids. The Long Duration Exposure Facility (LDEF) was a bus sized spacecraft. It was returned after 5.7 years in low Earth orbit. The LDEF was placed in low Earth orbit (LEO) by the space shuttle Challenger in April 1984 and retrieved by the space shuttle Columbia in January 1990. On the LDEF over 30 000 impacts were found (these craters were visible to the naked eye and larger than 0.5 mm). Form that sample about 1000 were chemically analyzed in order to investigate the origin of the projectiles. The largest crater found on LDEF had a diameter of 5 mm and was probably caused by a particle of 1 mm. Some impacts were clustered in time. On the European Retrievable Carrier (EURECA), the largest impact crater diameter was 6.4 mm. The returned solar array of the HST (Hubble Space Telescope, NASA/ESA) had been the one with the highest orbit altitude. It was found that the impact flux for HST was considerably higher (factor 2-8) than for EURECA. The infra-red astronomical satellite (IRAS), launched in 1983 to perform a sky survey at wavelengths ranging from 8 to 120 μm was operational during 10 months near altitude of 900 km. 200 000 potential debris sightings are stored in a database. About 10 000 sightings are attributed to real objects. A plot of debris flux in low Earth orbit as a function of object size (cm) is given in Fig. 12.2 where the coordinates are logarithmic.

- analysis of returned spacecraft surface; Critical surfaces, such as the windows, on the Space Shuttle are examined after every flight.

Donald H. Humes and William H. Kinard from NASA Langley Research Center examined the WF/PC-I radiator with a microscope to measure the damage done by meteoroids and man-made orbital debris during its 3.6 years

12.3. ESA SPACE DEBRIS AND METEOROID MODEL

Figure 12.3: LDEF Retrieval off the coast of Baja California, Courtesy NASA

Table 12.1: Some examples of retrieved spacecraft and surfaces

Name	Orbit	In orbit	Exposed area
Salyut 4,6	350 km	1974-1979	$\sim 7\,\text{m}^2$
STS-7 Window (NASA)	295-320 km	June 1983	$\sim 2.5\,\text{m}^2$
SMM (NASA)	500-570 km	1980-1994	$2.3\,\text{m}^2$
LDEF (NASA)	340-470 km	1984-1990	$151\,\text{m}^2$
EURECA (ESA)	520 km	1992-1993	$35\,\text{m}^2$
HST (solar array)	610 km	1990-1993	$62\,\text{m}^2$
Mir	390 km	1986-1998	$\sim 15\,\text{m}^2$

in orbit. They measured about 100 possible impact sites and rated them by size on an arbitrary scale of 1 to 10 (10 being the largest). They found 14 impact craters with a diameter greater than 450 microns.

At NASA a hypervelocity impact technology facility is under operation (HITF).

12.3 ESA Space Debris and Meteoroid Model

The four main activities of ESA-ESOC space debris task group are:

- development of a meteoroid and debris reference model;

- radar measurements of mid-size debris; these are necessary since current models in low earth orbit suffer from significant uncertainties about objects smaller than about 50 cm. This is essential for spacecraft which require protection; it is currently technically not feasible to shield against objects larger than 1 cm. The feasibility of detecting and tracking medium-size debris (1 to 50 cm) with a high power radar at the Forschungsgesellschaft für Angewandte Naturwissenschaften (FGAN) in Germany was investigated.

- Optical measurements; these are suited for objects in high altitude orbits. The detectors use CCD and a 1 m telescope will be operated by ESA at the Teide observatory in Tenerife.

- Analysis of spacecraft surfaces returned from space

The main aim of the mathematical model is a description of the debris and meteoroid environment at altitudes between low Earth orbit (LEO) and the geostationary orbit (GEO). The minimum size of an object is 0.1 mm. The model is based on the catalogued population and on known break-ups of spacecraft and rocket upper-stages in orbit. The initial distribution of fragments is described in terms of their position, velocity, mass. The objects are then propagated forward in time taking into account the relevant perturbations.

12.4 Detection of Space Debris

Remote sensing of space debris from ground-based measurements falls into two categories:

- Radar measurements: these have been used for space debris in low Earth orbit (LEO).

- Optical measurements: these have been used for high Earth orbit (HEO). For passive optical measurements the intensity of the signal from space debris is inversely proportional to the square of its distance or altitude:

$$I_{\text{optical}} \sim 1/r^2 \tag{12.1}$$

The incident illumination from the Sun is essentially independent of altitude. For radar measurements:

$$I_{\text{radar}} \sim 1/r^4 \tag{12.2}$$

since radars must provide their own illumination. Therefore, optical telescopes of modest size are more suitable than most radars for detection of debris at high altitudes. On the other hand, radars are better suited to detect objects in LEO.

12.4.1 Radar Measurements

Ground-based radars are well suited to observe space objects because

- all weather,
- all day-and-night

12.4. DETECTION OF SPACE DEBRIS

performance. There are two types for space object measurements:

1. Radars with mechanically controlled beam direction using parabolic reflector antennas; here, only objects in the field of view (which is given by the mechanical direction of the parabolic reflector antenna) can be observed; used for tracking or imaging satellites.

2. Radars with electronically controlled beam direction using phased array antennas. In that case multiple objects at different directions can be detected and measured simultaneously; used for tracking and search tasks.

In the tracking mode the radar follows an object for a few minutes gaining data on angular direction, range, range rate, amplitude and phase of the radar echoes. From these parameters the orbital elements can be derived In the beam-park mode, the antenna is kept fixed in a given direction and echoes are received from objects passing within its field of view. This yields statistical information on the number and size of detected objects; the determination of the orbit is less precise. There is also a mixed mode. From the radar measurements the following parameters can be derived:

1. orbital elements; thus the motion of the object's center of mass around Earth is defined.

2. Attitude; describes the motion of the object around its center of mass.

3. Size and Shape of the object.

4. Ballistic coefficient; this describes the rate at which the orbital semi-major axis decays.

5. Object mass,

6. material properties.

The main source of data for space debris in the size range of 1-30 cm is the NASA Haystake radar facility operated by MIT Lincoln Laboratory. Under an agreement with the US Air Force since 1990 data are collected. From the data information about the size, altitude and inclination of the space debris can be obtained. The data indicate that there are about 100 000 fragments in orbits with sizes down to 1 cm.

Another possibility to estimate the reentry of space debris (man made or meteoroids) is to use ionization radar measurements (operates at 50 MHz). Ionization radars detect the ionization trail behind reentering bodies. With such facilities one can detect meteors as small as 100 microns.

Radar measurements of space debris have been done at Haystack (US) and Goldstone radars (US), Russia and by Germany using the Research Establishment for Applied Science (FGAN) radar and the Effelsberg radio telescope. Haystack and Goldstone radars have provided a statistical picture of LEO debris at sizes down to 0.5 cm which was confirmed by FGAN. These measurements have proven that the debris population exceeds the natural meteoroid population for all sizes (except between 30 m and 500 m.

12.4.2 Telescopes

Space debris can be categorized into objects that reflect radar well but sunlight poorly. The other group reflects sunlight well but radar poorly. Thus, radar and optical telescopes see somewhat different debris populations. With the use of optical telescopes, debris at very high altitudes (e.g. in geosynchronous orbits, GEO) can be detected.

The US Space Command employs aperture telescopes of 1 m to track HEO objects. With these telescopes objects of 1 m at geosynchronous altitudes, corresponding to a limiting stellar magnitude of 16 can be detected. A limiting stellar magnitude of 17 or greater is needed to detect debris smaller than 1 m near GEO.

Most objects in GEO are intact; in 1978 a Russian Ekran satellite in GEO was observed to explode.

NASA is using two optical telescopes for measuring orbital debris: a 3 m diameter liquid mirror telescope which is referred to as the LMT, and a charged coupled device-equipped 0.3 m Schmidt camera, which is commonly referred to as the CCD Debris Telescope or CDT. The LMT consists of a 3 m diameter parabolic dish that holds four gallons of liquid mercury. The dish is spun up to a rate of 10 revolutions per minute. Centrifugal force and gravity cause the mercury to spread out in a thin layer over the dish creating a reflective parabolic surface that is as good as many polished glass mirrors.

12.4.3 Catalogues

There are two catalogues of space objects that are frequently updated:

- United States Space Command catalogue,
- Space Object catalogue of the Russian Federation.

Based on those two catalogues data are also archived in the Database and Information System Characterizing Objects in Space (DISCOS) of ESA. The National Space Development Agency (NASDA) of Japan is studying a debris database. Current catalogues contain information on satellites and debris as small as 10-30 cm in diameter. Some recent activities are aimed to provide detection of 5 cm objects at altitudes below 600 km. For smaller sizes modelers must use statistical measurements.

12.4.4 Risk Assessments

Risk assessments are utilized in the design of manned and unmanned spacecraft. They aid in the placement and protective shielding design. This is of course only feasible for critical subsystems and components. It becomes extremely important in the system design of large communication satellite constellations. In Table 12.2 a summary of the studies made so far is given.

For GEO the situation is more complicated. The number of space debris of less than 1 m in diameter is not well known. Moreover, there is no natural removal

Table 12.2: Mean time between impacts on a satellite with a cross-section area of 10 m^2

Height of circular orbit	Objects 0.1-1.0 cm	Objects 1-10 cm	Objects >10 cm
500 km	10-100 yrs	3 500-7 000 yrs	150 000 yrs
1 000 km	3-30 yrs	700 - 1 400 yrs	20 000 yrs
1 500 km	7-70 yrs	1 000-2 000 yrs	30 000 yrs

mechanism for satellites in GEO. One can estimate an annual collision probability for an average operational satellite with other catalogued objects at 10^{-5}.

Another problem concerns the re-entry. Since the last 40 years 16 000 known re-entries of catalogued space objects are known. No significant damage or injury occurred which can be attributed to the large expanse of ocean surface and sparse population density in many land regions. During the past years, approximately once each week an object with a cross section of 1 m^2 or more entered the Earth's atmosphere. The risk of re-entry comes from:

- Mechanical impact,

- chemical contamination,

- radiological contamination.

Since about 12% of the present catalogued space debris population consists of objects discarded during normal satellite deployment (fasteners, yaw, weights, nozzle covers, lens caps, tethers,...) one should take mitigation measures against these objects. 85% of all space debris larger than 5 cm result from fragmentation of upper stages. In 1996 the French CERISE spacecraft was struck and partially disabled by an impact fragment which most probably came from an exploded Ariane upper stage.

12.4.5 Shielding

Protection against particles 0.1-1 cm size can be achieved by shielding spacecraft structures. Objects 1-10 cm in size cannot be shielded nor can they be routinely tracked by surveillance networks. Protection against these particles can be achieved through special features in the design (e.g. redundant systems, frangible structures...). Physical protection against particles larger 10 cm is not technically feasible. In front of the spacecraft wall single sheet Whipple bumpers or complex layers of metal and ceramic/polymer fabrics can be used for shieldings. They break up the impacting particle and absorb the energy of the resulting ejecta. Bumper shields should be positioned at a sufficient distance from the shielded object.

The penetration depth (damage potential) of an impacting object depends on:

- mass,

- velocity,

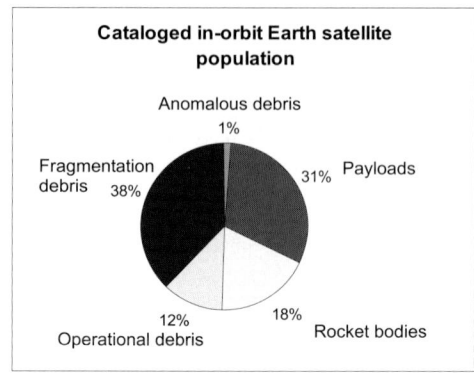

Figure 12.4: Segments of the cataloged in-orbit Earth satellite population.

- shape of the object; and of course

- material properties of the shield.

There are several models (NASA: BUMPER, ESA: ESABASE, Russia: BUFFER and COLLO).

For manned spacecraft shield designs offer protection against objects smaller than 1 cm. The PNP (probability of penetration) is an important criterion for shield design. One can also install automatic detection systems to locate damage. For EVA (extravehicular activities) current spacesuits have many features with inherent shielding qualities to offer protection from objects of sizes up to 0.1 mm. By properly orientating their spacecraft, astronauts may also be able to use their vehicles against the majority of space debris or direct meteoroid streams. The United States Space Surveillance Network (SSN) and the Russian Space Surveillance System(SSS) monitor the LEO environment to warn crewed spacecraft if an object is projected to approach within a few km. If an object is predicted to pass through a box of $5 \times 225 \times 5$ km oriented along the flight path of the United States Space Shuttle, the SSN sensor intensifies its tracking of the potential risk object. If the improved fly-by prediction indicates a conjunction within a box of $2 \times 5 \times 2$ km an avoidance manoeuvre is performed. During 1986-1997 4 such evasive manoeuvres were executed. Collision avoidance manoeuvres were performed by the ESA satellite ERS-1 in June 1997 and March 1998 and by the CNES satellite SPOT-2 in July 1997.

Calculations made prior to the launch of spacecrafts permit the establishment of safe launch windows.

For unmanned spacecraft, lower PNPs are tolerable.

12.5 Literature

An overview of fragmentations of LEO Upper Stages was given by Chernyavskiy et al. (1994).

A technical report on space debris was given from the Scientific and Technical Subcommittee of the United Nations Committee on the Peaceful uses of Outer Space (1999).

Chapter 13

Appendix

13.1 Bibliography

Adams L., Harboe-Sorensen R., Daly E. and Ward J., 1989, IEEE Trans. Nucl. Sci. N-S36, 6, 2339 (1989).

Adams J.H., Jr., 1986, NRL Memorandum Report 5901 (1986).

Alley, R.B., Clark, P. U., 1999, Annual Review Of Earth And Planetary Sciences, Volume 27, 149

Alvarez, L.W., Alvaraez, W., Asaro, F., Michel, H.V., 1980, Science, 208, 1095

Andersson, L., Eliasson, L., Wintoft, P., 1999, in Proccedings of the Workshop on Space Weather, 11-13 November 1998, ESTEC, Noordwijk, The Netherlands, WPP-155

Angell, J.K., Korshover, J., 1976, Month Weather Rev., 101, 426-443

Arnaud, J., Mein, P., Rayrole, J., 1998, in Crossroads for European Solar and Heliospheric Physics. Recent Achievements and Future Mission Possibilities, Puerto de la Cruz, Tenerife, Canary Islands, Spain. March 23-27,1998. Organised by ESA Solar Physics Planning Group (SPPG) and Instituto de Astrofisica de Canarias (IAC). European Space Agency, 1998., 213

Arpesella, C., et al. BOREXINO Prop. Vols. 1,2 ed. R. Raghavan et al., Univ. of Milano, Milano, 1992

Aschwanden, Markus J.; Poland, Arthur I.; Rabin, Douglas M., 2001 Annual Review of Astronomy and Astrophysics, 39, 175

Badhwar, G.D., 1997, Radiat. Res., 148, S3

Bahcall, J. N.; Pinsonneault, M. H., 1992, Rev. Mod. Phys., Volume 64, 885

Bahcall, J.N., Davis, R., Jr. , 2000 PASP, 112, Issue 770, 429

Bahcall, J.N., in Proc. of the 2nd int. workshop on Low Energy Solar Neutrinos, Univ. of Tokyo, Tokyo, Japan, World Scientific

Bahcall, J.N., Pinonncault, M.H., Basu, S., Christensen-Dalsgaard, J., 1997, Phys. Rev. Lett. 78, 171

Baliunas, S., Soon, W.H., 1995, Astrophys. J., 450, 896

Barnes J.R., Collier Cameron A., James D.J., Donati J.F. , 2001 MNRAS 324, 231

Basher, R. E. Basic science of solar radiation and its ultraviolet components, Proceedings of the seminar on solar ultraviolet radiation, R. Basher (Ed), NZMS, 1981.

Basu, D., 1992, Solar Physics, 142, 205

Basu, D., 1998, Sol. Phys. 183, 291

Basu, Sarbani, Christensen-Dalsgaard, J., Schou, J., Thompson, M. J., Tomczyk, S., 1996, Astrophysical Journal, 460, 1064

Beckers, J.M., and Brown, T.M. 1978, Oss. Mem. d. Oss. Astrofis. d. Arcetri, 106, 189

Beer, J., 2000, Space Science Rev., 94, 53

Benz A. O., 1993 Plasma Astrophysics, Kinetic Processes in Solar and Stellar

Bradley, P.D., Normand, E. , 1998, , IEEE Trans. Nucl. Sci., 45, no. 6 Coronae (Dordrecht: Kluwer)

Berger, A., 1980, Vistas in Astron., 24(2),103-122

Berner, R.A., Lasaga, A.C., Garrels, R.M., 1983, Amer. J. Sci. 283,641-683

Berry, P.A.M., Vistas in Astron., 30, 97-108

Birn, J., Hones, E. W., Jr., 1981, Journal of Geophysical Research, 86, 6802

Boberg, F., Wintoft, P., Lundstedt, H., 2000, Physics and Chemistry of Earth, Vol. 25, 4

Bogdan, T.J.,2000, Solar Physics, 192, 373

Bohor, B.F., 1990, in Global catastrophes in Earth history, V.L. Sharpton, P.D. Ward Eds., 335, Geolo. Society of America, Special Paper 247

Bochsler, P., 2001, Space Science Review, 97, 113

Boteler, D.H., R.J. Pirjola and H. Nevanlinna, 1998 Advances in Space Research, 22, 17

Bourgeois, J., Hansen, T.A., Wiberg, P.L., Kauffman, E.G., 1988, Science 241,567

Burlaga, L. F., E. Sittler, F. Mariani, and R. Schwenn,1981, J. Geophys. Res., 86, 6673, 1981.

Burlaga, L.F., 2001, Planetary and Space Science, 49, 1619

Calvo, R. A., Ceccato, H. A., Piacentini, R. D., 1995, Astrophysical Journal, 444, 916 Publication Date:

Canuto, V.M., Levine, J.S., Augustsson, T.R., Imhoff, C.L., Nature, 296, 816-820

Catling, D.C., Zahnle, K.J., McKay, C.P., 2001, Science, 293, 839

Cattaneo, F., Lenz, D., Weiss, N., 2001, The Astrophysical Journal, 563, L91-L94

Cattaneo, F., in SCORe'96 : Solar Convection and Oscillations and their Relationship, Proceedings of a workshop, held in Aarhus, Denmark, May 27 - 31, 1996, Eds.: F.P. Pijpers, J. Christensen-Dalsgaard, and C.S. Rosenthal, Kluwer Academic Publishers (Astrophysics and Space Science Library Vol. 225),201

Cattaneo, F., 1999, in Motions in the Solar Atmosphere, Proceedings of the Summerschool and Workshop held at the Solar Observatory Kanzelhöhe, Kärnten, Austria, September 1-12, 1997, edited by Arnold Hanslmeier and Mauro Messerotti, Kluwer Academic Publishers, Dordrecht/Boston/London, Astrophysics and Space Science Library, v. 239, 1999, ISBN 0-7923-5507-5, 119

Cauzzi, G.; Falchi, A.; Falciani, R., 2000, Astronomy and Astrophysics, 357, 1093

Cecchini, S., Giacomelli, G., Hasegan, D., Mandrioli, G., Maris, O., Patrizii, L., Plaian, A., Popa, V., Stefanov, L., Valeanu, V., 2000, Astrophysics and Space Science, 273, 35

Chandrasekhar, S., 1961, Hydrodynamik and Hydromagnetic Stability, Oxford Univ. Press

Chapman, C.R., Morrison, D., 1994, Nature, 367, 33

Charbonneau, P., and MacGregor, K.B. 1996, The Astrophysical Journal, 473, L59

Chen, J., Space Science Reviews, 95, 165

Chernyavskiy, G., Morozov, N., Johnson, N., McKnight, D., Maclay, T.,1994, IAA-94-IAA6.5.696

Choudhuri A.R., Auffret H. Priest, E.R., 1993, Sol. Phys. 143,39

Clancy, R.T., Rusch, D.W., J. Geophys. Res., 94, 3377-3393

Cockell, Ch.S., 2000, Origins of Life and Evolution of the Biosphere, 30, 467

Cook, E.R., D.M. Meko, and C.W. Stockton, 1997, Journal of Climate 10, 1343-1356.

Crowley, T.J. ,2000, Science, Volume 289, 270-277

Crowley, T.J. and G.R. North, Paleoclimatology - Oxford Monographs on Geology and Geophysics, Oxford University Press, Inc. New York, NY. 1991.

D'Arrigo, R.D. and Jacoby, G.C., 1993, Climatic Change 25: 163-177.

Dearborn, D.S.P., Blake, J.B., 1980, Astrophysical Journal, 237, 616

Dravins, D. , Lindegren, L. , Nordlund, A. , Vandenberg, D. A., 1993a, Astrophysical Journal, 403, 385

Dravins, D. , Linde, P. , Fredga, K. , Gahm, G. F. , 1993b, Astrophysical Journal, 403, 396

Duvall, T. L., Jr., Kosovichev, A. G., Scherrer, P. H., Bogart, R. S., Bush, R. I., de Forest, C., Hoeksema, J. T., Schou, J., Saba, J. L. R., Tarbell, T. D., Title, A. M., Wolfson, C. J., Milford, P. N., 1997, Solar Physics, 170, 63

Dyer, C., 2001 in SOLSPA 2001, ESA SP Ser 477

Dziembowski, W.A., Goode, P. R., Kosovichev, A.J., Schou, J., 2000, Astrophysical Journal, 537, 1026

Eddy, J.A., 1976, Science, 192, 1189-1202

Eddy, J.A., 1988, Secular Solar and Geomagnetic Variations in the Last 10 000 years, eds. F.R. Stephenson and A.W. Wolfendale, Dordrecht, Kluwer, 1-23

Eddy, J.A., 1977, Climate Change, 1 173-190

Espagnet, O., Muller, R., Roudier, T., Mein, P., Mein, N., Malherbe, J. M. 1996, Astronomy and Astrophysics, 313, 297

Etheridge, D. M.; Steele, L. P.; Langenfelds, R. L.; Francey, R. J.; Barnola, J.-M.; Morgan, V. I., 1996, Journal of Geophysical Research, 101, Issue D2, 4115

Evans, L.G., Starr, R., Brückner, J., Boynton, W.V., Bailey, S.H., Trombka, J.L., 1999, Nuclear Instruments and Methods in Physics Research, Sec. A., Vol. 422, 586

Favorite, F., and McLain, D. R., 1973, Nature, 244, 139

Fisher, G. H.; Fan, Y.; Longcope, D. W.; Linton, M. G.; Pevtsov, A. A., 2000, Solar Physics, 192, 119

Fleck, B., 2001 The Dynamic Sun, Proceedings of the Summerschool and Workshop held at the Solar Observatory Kanzelhöhe, Kärnten, Austria, August 30 - September 10, 1999, edited by Arnold Hanslmeier, Mauro Messerotti and Astrid Veronig, Kluwer Academic Publishers, Dordrecht/Boston/London, Astrophysics and Space Science Library, v. 259, ISBN 0-7923-6915-7, 336 , 1

Foukal, P., and Lean, J., 1990, Science, 247, 556

Foukal, P.V., Lean, J., 1988, Astrophysical Journal, 328, 347-357

Friis-Christensen, E., Lassen, K., 1991, Science, 254, 698

Fröhlich, C., 2000, Space Science Review, 94,15

Emilio, M., Kuhn, J.R., Bush, R.I., Scherrer, P., 2000, Astrophysical Journal, 543, 1007-1010

Fröhlich, C., 1987, J. Geophys. Res., 92, 796-800

Fröhlich, C., Romero, J., Roth, H., Wehrli, C., Andersen, B. N., Appourchaux, T., Domingo, V., Telljohann, U., Berthomieu, B., Delache, P., Provost, J., Toutain, T., Crommelynck, D., Chevalier, A., Fichot, A., Däppen, W., Gough, D. O., Hoeksema, T., Jiménez, Gómez, M., Herreros, J., Roca-Cortés, T., Jones, A. R., Pap, J.and Willson, R. C.: 1995, Solar Physics, 162, 101

Gaizauskas V 1989 Preflare activity Solar Physics, 121, 135-52

Gerard, J.C., 1990, Phil. Trans. Roy. Soc., London, in Press

Gertsch, R., Remo, J.L., Sour-Gertsch, L., 1997, in Near Earth Objects, Ann. of the New York Academy of Sciences 822, 468

Gibson, S.E., Space Science Review, 97, 69

Gilliland, R.L., 1981, Astrophysical Journal, 248, 1144-1155

Gilliland, R.L., 1980, Astrophysical Journal, 248, 1144

Gleisner, H., Lundstedt, H.,2001, Journal of Geophysical Research, Volume 106, 8425

Goldreich, P., Murray, N., Kumar, P. 1994, Astrophysical Journal, 424, 466

Goode, P. R., Strous, L. H. Rimmele, Th. R., Stebbins, R. T. , 1998, Astrophysical Journal Letters 495, L27

Gough, D.O., 1981, Solar Physics, 74, 21-34

Gough, D. O., Kosovichev, A. G., Toomre, J., Anderson, E., Antia, H. M., Basu, S., Chaboyer, B., Chitre, S. M., Christensen-Dalsgaard, J., Dziembowski, W. A., Eff-Darwich, A., Elliott, J. R., Giles, P. M., Goode, P. R., Guzik, J. A., Harvey, J. W., Hill, F., Leibacher, J. W., Monteiro, M. J. P. F. G., Richard, O., Sekii, T., Shibahashi, H., Takata, M., Thompson, M. J., Vauclair, S., Vorontsov, S. V., 1996, Science, 272, 1296

Graham, G.A., Sexton, A., Grady, M.M., Wright, I.P.,1997, Advances of Space Research, 20, 1461-1465

Haigh, J.D., 1994, Nature, 370, 544-546

Haigh, 1996, Science, 272, 981

Haisch, B., Strong, K.T., Rodono, M., 1991, Ann. Rev. of Astron. and Astrophys., 29, 275

Hansen, J.E., Lacis, A.A., 1990, Nature, 346, 713

Hathaway, D. H.; Beck, J. G.; Bogart, R. S.; Bachmann, K. T.; Khatri, G.; Petitto, J. M.; Han, S.; Raymond, J. ,2000, Solar Physics, 193, 299

Haxton, W.C., 2001, in: Current aspects of neutrino physics. David O. Caldwell (ed.). Physics and astronomy online library. Berlin: Springer, ISBN 3-540-41002-3, 2001, p. 65 - 88

Heckman, G.R., 1988, in JPL, Cal. Inst. of Techn., Interplanetary Particle Environment, proc., N8928454-22-90, 91

Hernandez, J. V., Tajima, T. Horton, W., 1993, Geophysical Research Letters, 20, 23, 2707

Hills, J.G., Mader, Ch.,L., 1997, Near Earth Objects, Ann. New York Acad. of Sciences, Vol. 822, 381

Hines, C.O., 1974, Atmos. Sci., 31, 589-591

Hoekzema, N. M., Brandt, P. N., Rutten, R. J., 1998, Astronomy and Astrophysics, 333, 322

Hoeksema, J. T. ,Schou, J. , Saba, J. L. R. , Tarbell, T. D. , Title, A. M. , Wolfson, C. J. Milford, P. N. 1997, Solar Physics, 170, 63.

Holland, H.D., 1978, The Chemistry of the Atmosphere and the Oceans, New York, Wiley

Hood, L.L., J. Geophys. Res., 91; 5264-5276

Hood, L.L., J. Geophys. Res., 92, 876-888

Houghton, G.J. Jenkins, and J.J. Ephraums (eds.),Climate Change: The IPCC Scientific Assessment. Cambridge University Press, Cambridge, UK.

Hoyt, D.V., and K. H. Schatten, 1997. The Role of the Sun in Climate Change, Oxford University Press, 279 pp.

Hoyt, D.V., Schatten, K.H., 1993, J. Geophys. Res., 98, 18895

Hughes, M.K. and H.F. Diaz, 1994, Climatic Change, Volume 26, 109

Hurlburt, N.E., 1999, in SOHO-9 Workshop "Helioseismic Diagnostics of Solar Convection and Activity", Stanford, California, July 12-15

Ishii, T. T.; Kurokawa, H.; Takeuchi, T. T. , 2000, in The Sun and Space Weather, 24th meeting of the IAU, Manchester, England

Johnson, S.J., Dansgaard, W., Clausen, H.B., Langway, C.C., 1970, Nature 227, 482-483

Kalkofen, W., 1997, Astrophysical Journal Lett. 436,L145

Kalkofen, W., 1990, in Basic Plasma Processes on the Sun (A92-30901 12-92). Dordrecht, Netherlands, Kluwer Academic Publishers, 1990, 197

Kalkofen, W.; Keller, C. U.; Smartt, R.; Hasan, S. S., 2000, Astronomy and Astrophysics, 363, 279

Kasting, 1989, Paleogeog. Paleoclimat. Paleoecol., 75, 83-95

Kasting, J.F., Whittet, D.C.B., Sheldon, W.R., 1997, Origins of Life and Evolution of the Biosphere, v. 27, 413

Keating, G.M., Pitts, M.C., Brasseur, G., De Rudder, A., 1987, J. Geophys. Res. 92, 889-902

Kerr, R.A., 1988, Science, 242, 124-125

Kiefer, J., 2001, Phys. Med. 17 Suppl 1, 1

Khomenko, E.V., Kostik, R.I., Shchukina, N.G., 2001, Astronomy and Astrophysics, 369,660

Kneer, F., von Uexküll, M., 1999, in Motions in the Solar Atmosphere, Proceedings of the Summerschool and Workshop held at the Solar Observatory Kanzelhöhe, Kärnten, Austria, September 1-12, 1997, edited by Arnold Hanslmeier and Mauro Messerotti, Kluwer Academic Publishers, Dordrecht/Boston/London, Astrophysics and Space Science Library, v. 239, 1999, ISBN 0-7923-5507-5, 99

Kosovichev A. G. et al., 1998, in Sounding solar and stellar interiors, Proceedings of the 181st symposium of the International Astronomical Union, held in Nice, France, September 30 - October 3, edited by Janine Provost, Francois-Xavier Schmider. 1996 Publisher: Dordrecht Kluwer Academic Publishers, 1998 438 p.

Krause, F., and Rädler, K.H., 1980, Mean - field magnetohydrodynamics and dynamo theory, Pergamon, Oxford

Krieger, A. S., A. F. Timothy, and E. C. Roelof, 1973, Sol. Phys., 29, 505

Krymskii, A. M.; Breus, T. K.; Ness, N. F.; AcuÑa, M. H., 2000, Space Science Review, 92, 535

Kubo, Y., 1993, Publ. Astron. Soc. Japan, 45, 819

Labitzke, K., 1987, Geophys. Res. Lett., 14, 535-537

Labitzke, K., Van Loon, H., 1988, J. Atmos. Terrestr. Phys., 50, 197-206

Laclare, F., Delmas, C., Coin, J.P., Irbh, A., 1996, Solar Physics 166, 211

Lagache,M., 1976, Geochim. Cosmochim. Acta, 40,157-161

Laskar, J., Joutel, F., Boudin, F.: 1993, O Astronomy and Astrophysics, 270, 522

Lammer, H., Stumptner, W., Molina-Cuberos, G.J., Bauer, S.J., Owen, T.,2000, Planetary and Space Science, 48, 529

Lean, J., Beer, J., Bradley, R., 1995, Geophys. Res. Lett. 22, 3195

Lean, J., 1997, Ann. Rev. Astron. Astrophys., 35, 33

Lean, J.L.,2000, American Astron. Society, SPD meeting 32,08.03

Lehtinen, M. and R. Pirjola, 1985, Annales Geophysicae, 3, 479

Leighton, R.B., Noyes, R.W., Simon, G.W., 1962 Astrophysical Journal, 135, 474

Leister, N.V., 1990, Rev. Mex. Astron. Astrofis., 21, 567

Lepping, R. P.; Berdichevsky, D. B., 2001, American Geophysical Union, Spring Meeting 2001, abstract SH61A-07

Lewis, M.R., Carr, M., Feldman, G.C., Esaias, W., and McClain, C., 1990, Nature, 347, 543

Lites, B. W.; Rutten, R. J.; Berger, T. E., 1999, Astrophysical Journal, 517, 1013

Lockwood, M., Stamper, R., Wild, M.N., 1999, Nature, 399, 437

Lockwood, M., Foster, S., 2000, ESA-SP 463

Lovelock, J.E.,1988, The Ages of Gaia, New York, Norton

Lovelock, J.E., 1979, Gaia: A New Look at Life on Earth, Oxford Univ. Press

Lydon, T.J., Sofia, S., 1995, Astrophysical Journal Supplement, 101, 357

McBride, N., McDonnell, J.a.m., 1999, Planetary and Space Science, 47, 1005

Mc Cormac, B.N., Seliga, T.A., eds., 1979, Solar-Terrestrial influences on Weather and Climate, Dordrecht,: D. Reidel.

Mc Donald, A.B.,1991, in Proc. of the 9th Lake Luise Winter Inst., A. Astbury et al. (eds), World Scientific,1

McDonnell, J.A.M., Gardner, D.J., Icarus, 133, 25

McDougall, J.D., 1988, Science, 239, 485

McGregor, K.B., and Charbonneau, 1997, Astrophysical Journal, 486, 484

McKay, G.F., Dubeau, J., Thomson, I., 1993 in Proceedings of Spacebound 93 Conference, Canadian Space Agency, Ottawa, Canada, 11

McKenzie, R. L., and J. M. Elwood, Intensity of solar ultraviolet radiation and its implications for skin cancer,1990, NZ Medical Journal, 103, 152 1990.

McLean, D. J. and Labrum, N. R. ed 1985 Solar Radiophysics (Cambridge Univ. Press)

McIntosh,1990, Sol. Phys. 125, 251

Mann, M.E., Bradley, R.S., and Hughes, M.K., Geophys. Res. Lett., 1999, 26, 759

Markvart, T., Cumberbatch, T.J., Dollery, A.A., Walkden, M., Photon and electron degradation of boron-doped FZ silicon solar cells, Proc. Third European Symposium on Photovoltaic Generators in Space (1982), p.109

Markvart, T., Willoughby, A.F.W., Dollery, A.A., Radiation-resistant silicon solar cell, Proc. 19th IEEE Photovoltaic Specialists Conference (1987), p. 709 J.W. Peters, T.

Markvart, T., Willoughby, A.F.W., Goodbody, G.C., Dollery, A.A., Defect interactions in silicon solar cells, Second Workshop on Radiation-Induced and/or Process-Related Electrically Active Defects in Semiconductors-Insulator Systems, North Carolina, U.S.A. (1989).

Markvart, T., Radiation Damage in Solar Cells: a review, Journal of Material Science: Materials in Electronics, Vol. 1, No. 1 (1990)

Marsh, N. D., Svensmark, H., 2000, Physical Review Letters 85, 5004

Mein, P., Rayrole, J., 1997, in 1st Advances in Solar Physics Euroconference. Advances in Physics of Sunspots, ASP Conf. Ser. Vol. 118., Eds.: B. Schmieder, J.C. del Toro Iniesta and M. Vazquez, p. 334.

Melosh, H.J., Schneider, N.M., Zahnle, K.J., Latham, D., 1990, Nature, 343, 251

Morel, P., Provost, J., Berthomieu, G., 1997, Astronomy and Astrophysics, 327, 349

Moreno-Insertis, F. ,1994, in Solar Magnetic Fields, eds. M. Schüssler and W. Schmidt, (Cambridge: Cambridge Univ. Press), 117

Muller, R., 1999, in Motions in the Solar Atmosphere, Proceedings of the Summerschool and Workshop held at the Solar Observatory Kanzelhöhe, Kärnten, Austria, September 1-12, 1997, edited by Arnold Hanslmeier and Mauro Messerotti, Kluwer Academic Publishers, Dordrecht/Boston/London, Astrophysics and Space Science Library, v. 239, 1999, ISBN 0-7923-5507-5, p. 35-70

Narain, U.; Ulmschneider, P. , 1990, Space Science Reviews, vol. 54, 377

National Research Council, Space Studies Board, 2000, Radiation and the International Space Station: Recommendations to reduce risk. Washington, DC, National Academy Press

Neckel, H., 1995, Sol. Phy. 156,7

Neftel, A.; Moor, E.; Oeschger, H.; Stauffer, B. , 1985, Nature, 315, 45

Noel, F., 1997, Astronomy and Astrophysics, 325, 825

Normand, E., IEEE Trans. Nucl. Sci.,1996 43, 461

November, L.J., Toomre, J., Gebbie, K.B., Simon, G., 1981, Astrophysical Journal, 245, L123

Ribes, E., Ribes, J.C., Barthalot, R., Nature, 326, 52-55

Ribes, E., Beardsley, B., Brown, T.M.,Delach, Ph., Laclare, F., Leister, V.N., 1991, in The Sun in Time, ed. C.P. Sonett, M.S. Giampapa, M.S. Matthews (Tucson Ariz. Univ. Press), 59

Otto, A., 1995, Rev. Geophys. Vol. 33 Issue S1, 657

Palle B., E., Butler, C.J., 2000, ESA SP-463

Perry, C.A., 1994, International Journal of Climatology, 14, 969

Peters, J.W., Markvart, T., Willoughby, A.F.W., 1992, A study of radiation-induced defects in silicon solar cells showing improved radiation resistance, Materials Science Forum 83-87 1539

Pirjola, R., Boteler, D., Viljanen, A. and Amm, O., 2000 , Advances in Space Research, 26, 514.

Potgieter, M.S., 1998, Space Science Rev. 83, 147

Powell, J., Maise, G., Ludewig, H., Todosow, M., 1997, in Near Earth Objects, Ann. of the New York Academy of Sciences 822, 447

Priest, E., 2000, in The Sun and Space Weather, 24th meeting of the IAU, Manchester, England

Raup, D., 1989, Phil. Transactions of the Royal Society of London..., 325, 421

Reedy, R.C., Second Conference on the High Energy Radiation Background in Space, Snowmass, Co, 22-23 July 1997, 41

Reid, G.C., 1991, J. Geophys. Res. 96, 2835

Riklis, E., Emerit, I., Setlow, R.B.,1996, Advances in Space Research, Vol. 18, 51

Rieutord, M.; Roudier, T.; Malherbe, J. M.; Rincon, F., 2000, Astronomy and Astrophysics, 357, 1063

Roald, Colin B.; Sturrock, P. A.; Wolfson, Richard, 2000, Astrophysical Journal, 538, 960

Roudier, T., Malherbe, J. M., November, L., Vigneau, J., Coupinot, G., Lafon, M., Muller, R., 1997, Astronomy and Astrophysics, 320, 605.

Roxburgh, I.W., 1998, Astrophysics and Space Science, 261, 57

Ruderman, M.A., Chamberlain, J.W., Planet. Space Sci., 1975, 23, 247-268

Russell, C.T., 2001, Planetary and Space Science, 49, 1005

Rutten, R. J., Hammerschlag, R. H., Bettonvil, F. M., Suetterlin, P., 2000, American Astronomical Society, SPD meeting 32, 02.107

Ryutova, M. P.; Tarbell, T. D., 2000, American Astronomical Society, SPD meeting 32, 01.41

Sanchez, M., Parra, F., Soler, M., Soto, R., 1995, Astronomy and Astrophysics Suppl. 110, 351

Sello, S., 2001, Astronomy and Astrophysics, 377, 312

Schrijver, C.J., Zwaan, C. , 2000, Solar and Stellar Magnetic Activity Cambridge U. Press, New York, 2000.

Schwartzmann, D.W., Volk, T., 1989, Nature, 340, 457-460

Seckmeyer, G., McKenzie, R.L., 1992, Nature, 359, 135

Shibata, K., 1996, Adv. Space Res 17, 9-18

Settecerri 2, M.J. Matney 2, J.C. Zhang 2 , Haystack Radar Measurements of the Orbital Debris Environment, 1990-1994, JSC - 27436

Sforza, P.M., Remo, J.L., 1997, in Near Earth Objects, Ann. of the New York Academy of Sciences 822, 432

Simpson, J.A., 1998, Space Science Rev., 83, 7

Shinagawa, H., 2000, Advances in Space Research, 26, 1599

Shoemaker, E.M., Wolfe, R.F., Shoemaker, C.S., 1990, Geol. Soc. of America, Spec. Paper 247, 155

Snodgrass, H. B., 2001, American Astronomical Society Meeting 198, 71.02

Sobotka, M., 1999, in Motions in the Solar Atmosphere, Proceedings of the Summerschool and Workshop held at the Solar Observatory Kanzelhöhe, Kärnten, Austria, September 1-12, 1997, edited by Arnold Hanslmeier and Mauro Messerotti, Kluwer Academic Publishers, Dordrecht/Boston/London, Astrophysics and Space Science Library, v. 239, 1999, ISBN 0-7923-5507-5, , 71

Sofia, S., Okeefe, J., Lesh, J.R., Endal, A.S., 1979, Science, 204, L306

Sonett, C.P., Williams, G.E., 1985, J. Geophys. Res., 90, 12019-12026

Sonett, C.P., 1982, Geophys. Res. Let., 9, 1313-1316

Sofia, S., Heaps, W., Twigg, L.W., 1994, ApJ, 427, 1048

Song, Q., Cao, W., 1999, in Obs. Astrophys. in Asia and its Future, Ed. P.S. Chen, Yunan Obs. Chin. Acadm. Sc., 1999, 139

Solanki, S.K., 1997, 1st Advances in Solar Physics Euroconference. Advances in Physics of Sunspots, ASP Conf. Ser. Vol. 118., Eds.: B. Schmieder, J.C. del Toro Iniesta, M. Vazquez, p. 178

Solanki, S.K., Fligge, M., 2000, ESA SP-463, 51

Spruit, H.C., 1982, Astronomy and Astrophysics, 108, 348

Stauffer, B., 2000, Space Science Reviews, 94, 321

Stein, Robert F.; Nordlund, Å , 2000, Solar Physics, 192, 91

Strassmeier K.G. and Linsky J.L., 1996, Stellar surface structure, conference proceedings of IAU symposium 176, Kluwer Academic Publishers, Dordrecht, The Netherlands

Strous, L. H., Goode, P. R., Rimmele, Th. R., 2000, Astrophysical Journal, 535, 1000

Suzuki, Y., 1998, Space Science Reviews, 85, 91

Svensmark, H., Friis-Christensen, E., 1997, Journal of Atmospheric Solar-Terrestrial Physics 59, 1225

13.1. BIBLIOGRAPHY

Svensmark, H., 1998, Physical Review Letters 81, 5027 - 5030

Svenonius, B., Olausson, E., 1979, Paleogeo., Paleoclim., Paleoecol., 26, 89-97

Svensmark, H., 1998, Phys. Rev. Lett. 81, 5027

Svensmark, H. and Friis-Christensen, E., 1997, Atmosph. Terr. Phys. 59, 1225

Svestka Z, Jackson B V and Machado M E 1992 Eruptive Solar Flares (Berlin: Springer)

Tada, H.Y., Carter, Jr., J,R., Anspaugh, B.E., Downing, R.G., The Solar Cell Radiation Handbook, 3rd Edition, NASA/JPL Publ. 82-69 (1 Nov 1982)

Tajika, E., 2001, Earth and Planetary Science Letters, 160, 695

Takata, M., 1993, in Frontier of Neutrino Astrophysics, ed. by S. Suzuki and K. Nakamura, Univ. Acad. Press, Tokyo, 1993, 147

Takata, M., Shibahashi, H., 1998, Astrophysical Journal, 504, 1035

Technical Report on Space Debris, United Nations Committee on the Peaceful uses of Outer Space, UN, New York, 1999

Thomson, I., 1999, Mutat. Res., 430(2), 203

Totsuka, Y, 1996, in Proc. 18th Texas Symp. on Rel. Astrophys., Chicago, A.Olinto, J. Frieman and D. Schramm (eds.),World Scientific, Singapore

Treu, M.H., Worden, S.P., Bedard, M.G., Bartlett, R.K., 1998, Earth, Moon and Planets,82/83, 27

Turck-Chièeze, S., Lopes, I., 1993 Astrophysical Journal, 408, 347

Turck-Chièze, S., Couvidat, S, Kosovichev, A. G., Gabriel, A. H., Berthomieu, G., Brun, A. S., Christensen-Dalsgaard, J., Garca, R. A., Gough, D. O., Provost, J., Roca-Cortes, T., Roxburgh, I. W., Ulrich, R. K., Astrophysical Journal, 2001, 555, L69

Ulrich, R.K., Bertello, L., 1995, Nature, 377, 214

Ulmschneider, P., Priest, E. R., Rosner, R., 1991, in Mechanisms of Chromospheric and Coronal Heating, Proceedings of the International Conference, Heidelberg, 5-8 June 1990, XV, 649 pp. Springer-Verlag Berlin Heidelberg New York.

v.d.Lühe, O., 2001 The Dynamic Sun, Proceedings of the Summerschool and Workshop held at the Solar Observatory Kanzelhöhe, Kärnten, Austria, August 30 - September 10, 1999, edited by Arnold Hanslmeier, Mauro Messerotti and Astrid Veronig,
Kluwer Academic Publishers, Dordrecht/Boston/London,
Astrophysics and Space Science Library, v. 259, ISBN 0-7923-6915-7, 43

Walker, J.C.G., Hays, P.B., Kasting, J.F., 1981, J. Geophys. Res. 86, 9776-9782

Walsh, R.W., 2001 The Dynamic Sun, Proceedings of the Summerschool and Workshop held at the Solar Observatory Kanzelhöhe, Kärnten, Austria, August 30 - September 10, 1999, edited by Arnold Hanslmeier, Mauro Messerotti and Astrid Veronig,
Kluwer Academic Publishers, Dordrecht/Boston/London,
Astrophysics and Space Science Library, v. 259, ISBN 0-7923-6915-7, 336

Watanabe, S., Shibahashi, H., 2001, Publications of the Astronomical Society of Japan, vol.53, no. 3, p. 565

Walther, G., 1999, Astrophysical Journal, 513, 990

Willson, R. C., 1981, Solar Physics, 74, 217-229.

Willson, R. C., 1984, Space Science Reviews, 38, 203-242.

Willson, R.C., Hudson, H., Nature, 332, 810-812

Willson, R.C., Hudson, H.S., Fröhlich, C., Brusa, R.W., Science, 1987, 234, 1114-1117

Wittmann, A.D., Alge, E., Bianda, M., 1993, Solar Physics, 145, 205

Wu, J.-G., Lundstedt, H., Eliasson, L., Hilgers, A., 1999, in Proccedings of the Workshop on Space Weather, 11-13 November 1998, ESTEC, Noordwijk, The Netherlands, WPP-155

Yang T.C., Mei, M., George, K.A., Craise, L.M., 1996, Advances of Space Research, 18, 149

Zahnle, K.J., Walker, J.C. G., 1992, Rev. Geophys. Space Phys., 20, 280-292

Ziegler, J.F., Lanford, W.A., 1979, Science, 20, 776

Zwickl, R. D.; Doggett, K. A.; Sahm, S.; Barrett, W. P.; Grubb, R. N.; Detman, T. R.; Raben, V. J.; Smith, C. W.; Riley, P.; Gold, R. E.; Mewaldt, R. A.; Maruyama, T., 1998, Space Science Rev., 86, 633

13.2 Internet

Today's space weather can be found under:

http://www.sel.noaa.gov/today.html

The web site of the National Oceanic and Atmospheric Administration:

http://www.sel.noaa.gov/

ESA Space Weather Site:

13.2. INTERNET

http://www.estec.esa.nl/wmwww/spweather/

NASA Space weather resources:

http://spdf.gsfc.nasa.gov/space_weather/Space_Weather_at_SSDOO.html

Space Science Institute/ NASA and NSF site:

http://www.spacescience.org/SWOP/

Lund Space Weather Center

http://www.irfl.lu.se/

further references can be found in these sites.

List of Tables

2.1	Central wavelengths and bandwidth of the UBVRI filter set	7
2.2	B-V colors and effective temperatures of some stars	8
2.3	Spectral classification of stars	9
2.4	Effective Temperature as a function of spectral type	9
2.5	The principal reaction of the pp chain	13
2.6	Solar model: variation of temperature, luminosity and fusion rate throughout the Sun	15
4.1	Prominent chromospheric emission lines	43
4.2	Optical classification scheme of solar flares	52
4.3	Soft x-ray classification scheme of solar flares	52
4.4	Radio classification scheme of solar flares	53
4.5	Solar Diameter Measurements	68
7.1	Composition of the Earth's atmosphere	112
7.2	Current Greenhouse Gas Concentrations and Other Components	114
7.3	Historical CO_2 record from the Siple Station Ice Core	115
7.4	Typical values for the albedo.	130
7.5	Effects of Solar Radiation at different wavelengths on the Middle and Upper Atmosphere	133
7.6	Satellite measurements of the solar constant	135
7.7	Exospheric temperature at solar maximum and minimum	143
7.8	Various influences on the climate	150
7.9	Causes of Global Warming of about 0.5 C, 1880-1997	151
8.1	Radiation related units	157
8.2	Radiation dose limits in mSv for astronauts	159
8.3	Total average annual radiation does in the US	159
8.4	Single dose effects	160
8.5	Common shielding materials	168
9.1	Variation of the ionosphere	170
9.2	Corrected magnetic latitudes of some cities	174
9.3	Extension of the auroral zone. The first values given is the magnetic latitude (Lat), the second the Kp index.	174
9.4	Transformation between the Kp and the Ap index	175

9.5	Transformation between the Ap and the Cp index	175
9.6	Transformation between the Cp and the C9 index	176
9.7	Summary of geomagnetic indices	177
9.8	Navigation systems	178
9.9	Fuels for RTG's	183
11.1	Extinction of marine species. The end of the stage is given in Myr.	208
12.1	Some examples of retrieved spacecraft and surfaces	215
12.2	Mean time between impacts on a satellite with a cross-section area of $10\,\mathrm{m}^2$	219

List of Figures

2.1   A typical spiral galaxy. From a distant galaxy our own system would appear similar, the Sun would be located in one of the spiral arms. . . .   4

2.2   Sketch of the Hertzsprung-Russell-diagram with evolutionary path of the Sun.   6

3.1   Big Bear Solar Observatory   18

3.2   Optical path scheme of a vacuum telescope (e.g. Kitt Peak or VTT, Tenerife). Below the ground level, a vertical spectrograph is located. The solar image can be observed at the top of the optical bank that is shown as a black box in the sketch.   19

3.3   Drawing of the SOHO solar observatory (ESA & NASA)   21

3.4   Propagation of a wave throughout the outer solar atmosphere. On the abscissa is the time, on the ordinate the frequency. Within 5 min the frequency drifts from 80 MHz to 40 MHz indicating the propagation to the higher corona. Courtesy: H. Aurass, Th. Mann, AIP.   23

4.1   Variation of electron temperature and electron density in the solar atmosphere   26

4.2   Spectroscopic observation of solar granulation. The entrance of a spectrograph slit covers different granular/intergranular areas. Line profiles emanating from granules are blueshifted because matter moves upwards and profiles from intergranular areas are redshifted because matter moves away from the observer. This is valid for solar granulation observed near the disk center.   29

4.3   Solar granulation and small network bright points   32

4.4   Large sunspot showing the dark central umbra and the filamentary penumbra. Outside the penumbra the granulation pattern is clearly seen . . .   35

4.5   Relative Sunspot number   40

4.6   Butterflydiagram illustrating the equatorward motion of spots during the activity cycle.   41

4.7   Profile of the CaII line   48

4.8   Coronal hole seen by the solar satellite YOHKOH   58

4.9   Comet Hale Bopp (1997); the fainter ion tail is clearly seen.   59

LIST OF FIGURES

5.1 Left: waves with low and high l; the low l modes are reflected deeper than the high l modes. Right: Explanation how the waves are reflected in the solar interior. The wavefront (normal to the propagation) is deflected since the sound velocity is higher in deeper layers ($c_2 > c_1$). 77

5.2 Examples of several modes 78

5.3 l-ν diagram from MDI high-cadence full disk data shows mode frequencies up to 10 mHz and l=1000. 80

5.4 This diagram shows the solar rotation rate inferred from two months of MDI Medium-l data as a function of radius at three latitudes, 0 degrees, 30 degrees, and 60 degrees. 83

5.5 MDI Dopplerimage; left: the rotation of the Sun is clearly seen; right: the rotation of the Sun was eliminated and therefore only velocities due to granulation and supergranulation are seen. 84

6.1 Looped magnetic field lines in the solar chromosphere and corona. Photo: NASA 88

6.2 Principle of magnetic reconnection 92

6.3 Illustration of the ω effect. The field lines are wraped around because of the differential rotation of the Sun 101

6.4 The MHD relation between flows and magnetic fields 102

6.5 a) Cutaway images of solar rotation showing a peak and a trough of the 0.72R variation, with black indicating slow rotation, grey intermediate, and white fast. b) Variations with time of the difference of the rotation rate from the temporal mean at two radii deep within the Sun, with the site at 0.72 $R_\odot$ located above the tachocline and that at 0.63 $R_\odot$ below it, both sampling speeding up and slowing down in the equatorial region. Results obtained from GONG data for two different inversions are shown with black symbols, those from MDI with red symbols. (Image courtesy NSF's National Solar Observatory) 107

7.1 Major elements of the climate system 113

7.2 Temperature anomaly clearly showing the Little Ice Age 118

7.3 Cretaceous climate and land/sea distribution; Image credit: Crowley and North, 1991 118

7.4 Upper curve: average insolation of 65 degrees northern latitude (Watts per one square meter of a horizontal atmosphere) in mid-July. As seen, it varies from some 390 to 490 W/m^2. Middle curve: Global temperature (Vostok ice core). Lower Curve: Greenland, GRIP core. Image courtesy: Jan Hollan 119

7.5 Vostok Ice core. Different depth can be attributed to different ages 120

7.6 Variation of global temperature over the last 150 000 years 121

7.7 World Carbon Dioxide Emissions (US Dept. of Energy) 122

7.8 Effective radiating temperature of the Earth as a function of planetary Albedo A for three different values of the solar constant, a) 982, b) 1088 (dotted), c) present value 1360 (dashed). 129

7.9 Solar irradiance measurements from satellites 135

LIST OF FIGURES

7.10 Three-dimensional rendering of the angular distribution of the excess irradiance emitted at 500 nm by the active region studied at two stages of its development, together with the magnetogram. A more uniform brightening of the facular region at the later stage is apparent (after Vicente Domingo). 137

7.11 Reconstructed precipitation in northern New Mexico. Courtesy: Henri D Grissino-Mayer 138

7.12 Temperatures derived from tree rings. Here the Maunder minimum is not seen whereas a cold period between 1830 to 1870) 139

7.13 Thermospheric temperature changes, a) low solar activity, $F_{10.7} = 80$, $A_p = 0$, b) high solar activity, $F_{10.7} = 200, A_p = 80$ 141

7.14 Penetration of different solar light waves resp. their induced particles in the atmosphere 142

7.15 Relationship of Northern hemisphere mean (NH) temperature reconstruction to estimates of three candidate forcings between 1610 and 1995. . . . 153

8.1 DNA damage caused by radiation 158
8.2 Correlation of the occurrence of solar proton events with solar activity cycle (indicated by the sunspot number) 162
8.3 Typical clear-sky UV indices over New Zealand and its surrounding region. Seasonal variations are larger at low latitudes (denoted by numbers). 164

9.1 The Earth's ionosphere 171
9.2 Van Allen Belts 172
9.3 Radio signal propagation in the ionosphere. 181
9.4 Single event upsets; spatial distribution of errors from the UoSAT-3 spacecraft in polar orbit; please note the South Atlantic Anomaly . . . 187
9.5 Satellite lifetimes 189
9.6 Forces acting on a satellite in a low circular orbit. 191

10.1 After a flare or coronal mass ejection erupts from the Sun's surface, major disturbances arrive with a range of time delays and a storm begins to build in the space surrounding the Earth. 194
10.2 Summary of space weather effect in the Earth's environment 199

11.1 Estimated frequency of impacts on the Earth from the present population of comets and asteroids and impact craters. 207
11.2 Height in the atmosphere at which half the kinetic energy of a stony meteoroid is dissipated. Note that asteroids with > 100 m hit the ground with most of their original kinetic energy. 209

12.1 GEO and LEO objects as a source of space debris 211
12.2 Approximate measured debris flux in low Earth orbit by object size (sketch) 214
12.3 LDEF Retrieval off the coast of Baja California, Courtesy NASA 215
12.4 Segments of the cataloged in-orbit Earth satellite population. 220

Previously published in Astrophysics and Space Science Library book series:

- **Volume 273**: <u>Lunar Gravimetry</u>
  Author: Rune Floberghagen
  Hardbound, ISBN 1-4020-0544-X, April 2002
- **Volume 271**: <u>Astronomy-inspired Atomic and Molecular Physics</u>
  Author: A.R.P. Rau
  Hardbound, ISBN 1-4020-0467-2, March 2002
- **Volume 269**: <u>Mechanics of Turbulence of Multicomponent Gases</u>
  Authors: Mikhail Ya. Marov, Aleksander V. Kolesnichenko
  Hardbound, ISBN 1-4020-0103-7, December 2001
- **Volume 268**: <u>Multielement System Design in Astronomy and Radio Science</u>
  Authors: Lazarus E. Kopilovich, Leonid G. Sodin
  Hardbound, ISBN 1-4020-0069-3, November 2001
- **Volume 267**: <u>The Nature of Unidentified Galactic High-Energy Gamma-Ray Sources</u>
  Editors: Alberto Carramiñana, Olaf Reimer, David J. Thompson
  Hardbound, ISBN 1-4020-0010-3, October 2001
- **Volume 266**: <u>Organizations and Strategies in Astronomy II</u>
  Editor: André Heck
  Hardbound, ISBN 0-7923-7172-0, October 2001
- **Volume 265**: <u>Post-AGB Objects as a Phase of Stellar Evolution</u>
  Editors: R. Szczerba, S.K. Górny
  Hardbound, ISBN 0-7923-7145-3, July 2001
- **Volume 264**: <u>The Influence of Binaries on Stellar Population Studies</u>
  Editor: Dany Vanbeveren
  Hardbound, ISBN 0-7923-7104-6, July 2001
- **Volume 262**: <u>Whistler Phenomena</u>
  <u>Short Impulse Propagation</u>
  Authors: Csaba Ferencz, Orsolya E. Ferencz, Dániel Hamar, János Lichtenberger
  Hardbound, ISBN 0-7923-6995-5, June 2001
- **Volume 261**: <u>Collisional Processes in the Solar System</u>
  Editors: Mikhail Ya. Marov, Hans Rickman
  Hardbound, ISBN 0-7923-6946-7, May 2001
- **Volume 260**: <u>Solar Cosmic Rays</u>
  Author: Leonty I. Miroshnichenko
  Hardbound, ISBN 0-7923-6928-9, May 2001
- **Volume 259**: <u>The Dynamic Sun</u>
  Editors: Arnold Hanslmeier, Mauro Messerotti, Astrid Veronig
  Hardbound, ISBN 0-7923-6915-7, May 2001
- **Volume 258**: <u>Electrohydrodynamics in Dusty and Dirty Plasmas</u>
  <u>Gravito-Electrodynamics and EHD</u>
  Author: Hiroshi Kikuchi
  Hardbound, ISBN 0-7923-6822-3, June 2001
- **Volume 257**: <u>Stellar Pulsation - Nonlinear Studies</u>
  Editors: Mine Takeuti, Dimitar D. Sasselov
  Hardbound, ISBN 0-7923-6818-5, March 2001

- **Volume 256: Organizations and Strategies in Astronomy**
  Editor: André Heck
  Hardbound, ISBN 0-7923-6671-9, November 2000
- **Volume 255: The Evolution of the Milky Way**
  Stars versus Clusters
  Editors: Francesca Matteucci, Franco Giovannelli
  Hardbound, ISBN 0-7923-6679-4, January 2001
- **Volume 254: Stellar Astrophysics**
  Editors: K.S. Cheng, Hoi Fung Chau, Kwing Lam Chan, Kam Ching Leung
  Hardbound, ISBN 0-7923-6659-X, November 2000
- **Volume 253: The Chemical Evolution of the Galaxy**
  Author: Francesca Matteucci
  Hardbound, ISBN 0-7923-6552-6, May 2001
- **Volume 252: Optical Detectors for Astronomy II**
  State-of-the-art at the Turn of the Millennium
  Editors: Paola Amico, James W. Beletic
  Hardbound, ISBN 0-7923-6536-4, December 2000
- **Volume 251: Cosmic Plasma Physics**
  Author: Boris V. Somov
  Hardbound, ISBN 0-7923-6512-7, September 2000
- **Volume 250: Information Handling in Astronomy**
  Editor: André Heck
  Hardbound, ISBN 0-7923-6494-5, October 2000
- **Volume 249: The Neutral Upper Atmosphere**
  Author: S.N. Ghosh
  Hardbound, ISBN 0-7923-6434-1, (in production)
- **Volume 247: Large Scale Structure Formation**
  Editors: Reza Mansouri, Robert Brandenberger
  Hardbound, ISBN 0-7923-6411-2, August 2000
- **Volume 246: The Legacy of J.C. Kapteyn**
  Studies on Kapteyn and the Development of Modern Astronomy
  Editors: Piet C. van der Kruit, Klaas van Berkel
  Hardbound, ISBN 0-7923-6393-0, August 2000
- **Volume 245: Waves in Dusty Space Plasmas**
  Author: Frank Verheest
  Hardbound, ISBN 0-7923-6232-2, April 2000
- **Volume 244: The Universe**
  Visions and Perspectives
  Editors: Naresh Dadhich, Ajit Kembhavi
  Hardbound, ISBN 0-7923-6210-1, August 2000
- **Volume 243: Solar Polarization**
  Editors: K.N. Nagendra, Jan Olof Stenflo
  Hardbound, ISBN 0-7923-5814-7, July 1999
- **Volume 242: Cosmic Perspectives in Space Physics**
  Author: Sukumar Biswas
  Hardbound, ISBN 0-7923-5813-9, June 2000
- **Volume 241: Millimeter-Wave Astronomy: Molecular Chemistry & Physics in Space**

Editors: W.F. Wall, Alberto Carramiñana, Luis Carrasco, P.F. Goldsmith
   Hardbound, ISBN 0-7923-5581-4, May 1999
- **Volume 240: Numerical Astrophysics**
   Editors: Shoken M. Miyama, Kohji Tomisaka, Tomoyuki Hanawa
   Hardbound, ISBN 0-7923-5566-0, March 1999
- **Volume 239: Motions in the Solar Atmosphere**
   Editors: Arnold Hanslmeier, Mauro Messerotti
   Hardbound, ISBN 0-7923-5507-5, February 1999
- **Volume 238: Substorms-4**
   Editors: S. Kokubun, Y. Kamide
   Hardbound, ISBN 0-7923-5465-6, March 1999
- **Volume 237: Post-Hipparcos Cosmic Candles**
   Editors: André Heck, Filippina Caputo
   Hardbound, ISBN 0-7923-5348-X, December 1998
- **Volume 236: Laboratory Astrophysics and Space Research**
   Editors: P. Ehrenfreund, C. Krafft, H. Kochan, V. Pirronello
   Hardbound, ISBN 0-7923-5338-2, December 1998
- **Volume 235: Astrophysical Plasmas and Fluids**
   Author: Vinod Krishan
   Hardbound, ISBN 0-7923-5312-9, January 1999
   Paperback, ISBN 0-7923-5490-7, January 1999
- **Volume 234: Observational Evidence for Black Holes in the Universe**
   Editor: Sandip K. Chakrabarti
   Hardbound, ISBN 0-7923-5298-X, November 1998
- **Volume 233: B[e] Stars**
   Editors: Anne Marie Hubert, Carlos Jaschek
   Hardbound, ISBN 0-7923-5208-4, September 1998
- **Volume 232: The Brightest Binaries**
   Authors: Dany Vanbeveren, W. van Rensbergen, C.W.H. de Loore
   Hardbound, ISBN 0-7923-5155-X, July 1998
- **Volume 231: The Evolving Universe**
   Selected Topics on Large-Scale Structure and on the Properties of Galaxies
   Editor: Donald Hamilton
   Hardbound, ISBN 0-7923-5074-X, July 1998
- **Volume 230: The Impact of Near-Infrared Sky Surveys on Galactic and Extragalactic Astronomy**
   Editor: N. Epchtein
   Hardbound, ISBN 0-7923-5025-1, June 1998
- **Volume 229: Observational Plasma Astrophysics: Five Years of Yohkoh and Beyond**
   Editors: Tetsuya Watanabe, Takeo Kosugi, Alphonse C. Sterling
   Hardbound, ISBN 0-7923-4985-7, March 1998
- **Volume 228: Optical Detectors for Astronomy**
   Editors: James W. Beletic, Paola Amico
   Hardbound, ISBN 0-7923-4925-3, April 1998
- **Volume 227: Solar System Ices**
   Editors: B. Schmitt, C. de Bergh, M. Festou
   Hardbound, ISBN 0-7923-4902-4, January 1998

- Volume 226: **Observational Cosmology with the New Radio Surveys**
  Editors: M.N. Bremer, N. Jackson, I. Pérez-Fournon
  Hardbound, ISBN 0-7923-4885-0, February 1998
- Volume 225: **SCORe'96: Solar Convection and Oscillations and their Relationship**
  Editors: F.P. Pijpers, Jørgen Christensen-Dalsgaard, C.S. Rosenthal
  Hardbound, ISBN 0-7923-4852-4, January 1998
- Volume 224: **Electronic Publishing for Physics and Astronomy**
  Editor: André Heck
  Hardbound, ISBN 0-7923-4820-6, September 1997
- Volume 223: **Visual Double Stars: Formation, Dynamics and Evolutionary Tracks**
  Editors: J.A. Docobo, A. Elipe, H. McAlister
  Hardbound, ISBN 0-7923-4793-5, November 1997
- Volume 222: **Remembering Edith Alice Müller**
  Editors: Immo Appenzeller, Yves Chmielewski, Jean-Claude Pecker, Ramiro de la Reza, Gustav Tammann, Patrick A. Wayman
  Hardbound, ISBN 0-7923-4789-7, February 1998
- Volume 220: **The Three Galileos: The Man, The Spacecraft, The Telescope**
  Editors: Cesare Barbieri, Jürgen H. Rahe†, Torrence V. Johnson, Anita M. Sohus
  Hardbound, ISBN 0-7923-4861-3, December 1997
- Volume 219: **The Interstellar Medium in Galaxies**
  Editor: J.M. van der Hulst
  Hardbound, ISBN 0-7923-4676-9, October 1997
- Volume 218: **Astronomical Time Series**
  Editors: Dan Maoz, Amiel Sternberg, Elia M. Leibowitz
  Hardbound, ISBN 0-7923-4706-4, August 1997
- Volume 217: **Nonequilibrium Processes in the Planetary and Cometary Atmospheres: Theory and Applications**
  Authors: Mikhail Ya. Marov, Valery I. Shematovich, Dmitry V. Bisikalo, Jean-Claude Gérard
  Hardbound, ISBN 0-7923-4686-6, September 1997
- Volume 216: **Magnetohydrodynamics in Binary Stars**
  Author: C.G. Campbell
  Hardbound, ISBN 0-7923-4606-8, August 1997
- Volume 215: **Infrared Space Interferometry: Astrophysics & the Study of Earth-like Planets**
  Editors: C. Eiroa, A. Alberdi, Harley A. Thronson Jr., T. de Graauw, C.J. Schalinski
  Hardbound, ISBN 0-7923-4598-3, July 1997
- Volume 214: **White Dwarfs**
  Editors: J. Isern, M. Hernanz, E. García-Berro
  Hardbound, ISBN 0-7923-4585-1, May 1997
- Volume 213: **The Letters and Papers of Jan Hendrik Oort as archived in the University Library, Leiden**
  Author: J.K. Katgert-Merkelijn
  Hardbound, ISBN 0-7923-4542-8, May 1997

- **Volume 212: Wide-Field Spectroscopy**
  Editors: E. Kontizas, M. Kontizas, D.H. Morgan, G.P. Vettolani
  Hardbound, ISBN 0-7923-4518-5, April 1997
- **Volume 211: Gravitation and Cosmology**
  Editors: Sanjeev Dhurandhar, Thanu Padmanabhan
  Hardbound, ISBN 0-7923-4478-2, April 1997
- **Volume 210: The Impact of Large Scale Near-IR Sky Surveys**
  Editors: F. Garzón, N. Epchtein, A. Omont, B. Burton, P. Persi
  Hardbound, ISBN 0-7923-4434-0, February 1997
- **Volume 209: New Extragalactic Perspectives in the New South Africa**
  Editors: David L. Block, J. Mayo Greenberg
  Hardbound, ISBN 0-7923-4223-2, October 1996
- **Volume 208: Cataclysmic Variables and Related Objects**
  Editors: A. Evans, Janet H. Wood
  Hardbound, ISBN 0-7923-4195-3, September 1996
- **Volume 207: The Westerbork Observatory, Continuing Adventure in Radio Astronomy**
  Editors: Ernst Raimond, René Genee
  Hardbound, ISBN 0-7923-4150-3, September 1996
- **Volume 206: Cold Gas at High Redshift**
  Editors: M.N. Bremer, P.P. van der Werf, H.J.A. Röttgering, C.L. Carilli
  Hardbound, ISBN 0-7923-4135-X, August 1996
- **Volume 205: Cataclysmic Variables**
  Editors: A. Bianchini, M. Della Valle, M. Orio
  Hardbound, ISBN 0-7923-3676-3, November 1995
- **Volume 204: Radiation in Astrophysical Plasmas**
  Author: V.V. Zheleznyakov
  Hardbound, ISBN 0-7923-3907-X, February 1996
- **Volume 203: Information & On-Line Data in Astronomy**
  Editors: Daniel Egret, Miguel A. Albrecht
  Hardbound, ISBN 0-7923-3659-3, September 1995
- **Volume 202: The Diffuse Interstellar Bands**
  Editors: A.G.G.M. Tielens, T.P. Snow
  Hardbound, ISBN 0-7923-3629-1, October 1995
- **Volume 201: Modulational Interactions in Plasmas**
  Authors: Sergey V. Vladimirov, Vadim N. Tsytovich, Sergey I. Popel, Fotekh Kh. Khakimov
  Hardbound, ISBN 0-7923-3487-6, June 1995
- **Volume 200: Polarization Spectroscopy of Ionized Gases**
  Authors: Sergei A. Kazantsev, Jean-Claude Henoux
  Hardbound, ISBN 0-7923-3474-4, June 1995
- **Volume 199: The Nature of Solar Prominences**
  Author: Einar Tandberg-Hanssen
  Hardbound, ISBN 0-7923-3374-8, February 1995
- **Volume 198: Magnetic Fields of Celestial Bodies**
  Author: Ye Shi-hui
  Hardbound, ISBN 0-7923-3028-5, July 1994

- **Volume 193: Dusty and Self-Gravitational Plasmas in Space**
  Authors: Pavel Bliokh, Victor Sinitsin, Victoria Yaroshenko
  Hardbound, ISBN 0-7923-3022-6, September 1995
- **Volume 191: Fundamentals of Cosmic Electrodynamics**
  Author: Boris V. Somov
  Hardbound, ISBN 0-7923-2919-8, July 1994
- **Volume 190: Infrared Astronomy with Arrays
  The Next Generation**
  Editor: Ian S. McLean
  Hardbound, ISBN 0-7923-2778-0, April 1994
- **Volume 189: Solar Magnetic Fields
  Polarized Radiation Diagnostics**
  Author: Jan Olof Stenflo
  Hardbound, ISBN 0-7923-2793-4, March 1994
- **Volume 188: The Environment and Evolution of Galaxies**
  Authors: J. Michael Shull, Harley A. Thronson Jr.
  Hardbound, ISBN 0-7923-2541-9, October 1993
  Paperback, ISBN 0-7923-2542-7, October 1993
- **Volume 187: Frontiers of Space and Ground–Based Astronomy
  The Astrophysics of the 21st Century**
  Editors: Willem Wamsteker, Malcolm S. Longair, Yoji Kondo
  Hardbound, ISBN 0-7923-2527-3, August 1994
- **Volume 186: Stellar Jets and Bipolar Outflows**
  Editors: L. Errico, Alberto A. Vittone
  Hardbound, ISBN 0-7923-2521-4, October 1993
- **Volume 185: Stability of Collisionless Stellar Systems
  Mechanisms for the Dynamical Structure of Galaxies**
  Author: P.L. Palmer
  Hardbound, ISBN 0-7923-2455-2, October 1994
- **Volume 184: Plasma Astrophysics
  Kinetic Processes in Solar and Stellar Coronae**
  Author: Arnold O. Benz
  Hardbound, ISBN 0-7923-2429-3, September 1993
- **Volume 183: Physics of Solar and Stellar Coronae:
  G.S. Vaiana Memorial Symposium**
  Editors: Jeffrey L. Linsky, Salvatore Serio
  Hardbound, ISBN 0-7923-2346-7, August 1993
- **Volume 182: Intelligent Information Retrieval: The Case of Astronomy and
  Related Space Science**
  Editors: André Heck, Fionn Murtagh
  Hardbound, ISBN 0-7923-2295-9, June 1993
- **Volume 181: Extraterrestrial Dust
  Laboratory Studies of Interplanetary Dust**
  Author: Kazuo Yamakoshi
  Hardbound, ISBN 0-7923-2294-0, February 1995
- **Volume 180: The Center, Bulge, and Disk of the Milky Way**
  Editor: Leo Blitz
  Hardbound, ISBN 0-7923-1913-3, August 1992

- **Volume 179: Structure and Evolution of Single and Binary Stars**
  Authors: C.W.H. de Loore, C. Doom
  Hardbound, ISBN 0-7923-1768-8, May 1992
  Paperback, ISBN 0-7923-1844-7, May 1992
- **Volume 178: Morphological and Physical Classification of Galaxies**
  Editors: G. Longo, M. Capaccioli, G. Busarello
  Hardbound, ISBN 0-7923-1712-2, May 1992
- **Volume 177: The Realm of Interacting Binary Stars**
  Editors: J. Sahade, G.E. McCluskey, Yoji Kondo
  Hardbound, ISBN 0-7923-1675-4, December 1992
- **Volume 176: The Andromeda Galaxy**
  Author: Paul Hodge
  Hardbound, ISBN 0-7923-1654-1, June 1992
- **Volume 175: Astronomical Photometry, A Guide**
  Authors: Christiaan Sterken, J. Manfroid
  Hardbound, ISBN 0-7923-1653-3, April 1992
  Paperback, ISBN 0-7923-1776-9, April 1992
- **Volume 174: Digitised Optical Sky Surveys**
  Editors: Harvey T. MacGillivray, Eve B. Thomson
  Hardbound, ISBN 0-7923-1642-8, March 1992
- **Volume 173: Origin and Evolution of Interplanetary Dust**
  Editors: A.C. Levasseur-Regourd, H. Hasegawa
  Hardbound, ISBN 0-7923-1365-8, March 1992
- **Volume 172: Physical Processes in Solar Flares**
  Author: Boris V. Somov
  Hardbound, ISBN 0-7923-1261-9, December 1991
- **Volume 171: Databases and On-line Data in Astronomy**
  Editors: Miguel A. Albrecht, Daniel Egret
  Hardbound, ISBN 0-7923-1247-3, May 1991
- **Volume 170: Astronomical Masers**
  Author: Moshe Elitzur
  Hardbound, ISBN 0-7923-1216-3, February 1992
  Paperback, ISBN 0-7923-1217-1, February 1992
- **Volume 169: Primordial Nucleosynthesis and Evolution of the Early Universe**
  Editors: Katsuhiko Sato, J. Audouze
  Hardbound, ISBN 0-7923-1193-0, August 1991

Missing volume numbers have not yet been published.
For further information about this book series we refer you to the following web site: http://www.wkap.nl/prod/s/ASSL

To contact the Publishing Editor for new book proposals:
Dr. Harry (J.J.) Blom: harry.blom@wkap.nl